Estuarine hydrography and sedimentation

A handbook

Estuarine and brackish-water sciences association handbooks

Estuarine hydrography and sedimentation

A handbook

Edited by
K. R. DYER
Institute of Oceanographic Sciences
Taunton, England

CAMBRIDGE UNIVERSITY PRESS
CAMBRIDGE
LONDON · NEW YORK · MELBOURNE

Published by the Syndics of the Cambridge University Press
The Pitt Building, Trumpington Street, Cambridge CB2 1RP
Bentley House, 200 Euston Road, London NW1 2DB
32 East 57th Street, New York, NY 10022, USA
296 Beaconsfield Parade, Middle Park, Melbourne 3206, Australia

First published 1979

Printed in Great Britain at the
University Press, Cambridge

ISBN 0 521 22435 7 hard covers
ISBN 0 521 29496 7 paperback

Library of Congress Cataloging in Publication Data

Main entry under title:

Estuarine hydrography and sedimentation.

(Estuarine and brackish-water sciences association handbooks)

Includes index.

1. Estuarine oceanography. 2. Estuarine
sediments. I. Dyer, Keith R. II. Series:
Estuarine and Brackish water Sciences Association.
Estuarine and Brackish water Sciences Association handbooks.
GC97.E78 551.4′609 78-67308
ISBN 0 521 22435 7 hard covers
ISBN 0 521 29496 7 paperback

Contents

Contributors

A. T. Buller, Shell Exploration and Production UK, Shell Centre, London, UK

B. D'Olier, Department of Geology and Geography, Sir John Cass School of Science and Technology, City of London Polytechnic, London, UK

K. R. Dyer, Institute of Oceanographic Sciences, Crossway, Taunton, UK

N. C. Glen, Tidal Branch, Ministry of Defence, Hydrographic Department, Taunton, UK

D. J. Hooper, St Mark's, Buttsway, Milverton, Somerset, UK

B. Kjerfve, Belle W. Baruch Institute of Marine Biology and Coastal Oceanography, University of South Carolina, Columbia, South Carolina 29208, USA

I. N. McCave, Department of Environmental Sciences, University of East Anglia, Norwich, UK

J. McManus, Department of Geology, The University of Dundee, Scotland

Preface

I have frequently been approached by research students for advice on techniques of analysis and ways of interpreting data in fields which are rather foreign to them, even though they are within the general area of estuarine studies. One example went as follows: 'I am a marine biologist. The animals I study live in sediment. How can I analyse the sediment and present the results in a manner useful in establishing any relationships between the animals and the sediment, and also which would be accepted and understood by a marine geologist?' It is mainly this sort of question that this book is aimed to answer. Even then the book is not meant to be a 'cookery book' giving the detailed procedures, but a source where techniques are outlined together with their advantages and limitations, with references to other publications for details. From this point of view it should also provide a reference book for advanced undergraduate and for postgraduate teaching courses.

This handbook is sponsored by the Estuarine and Brackish Waters Sciences Association and two companion volumes are in preparation; one on estuarine chemistry and the other on estuarine biology.

I would like to thank the authors for cooperating in producing the book, though unfortunately the contributions have spanned from 1976 into 1978. Nevertheless I am sure that the material has not dated significantly in that time. Also I wish to thank Mrs D. Croston and Mrs P. Sloper for typing efficiently the various drafts.

Taunton, May 1978 K. R. DYER

1

Estuaries and estuarine sedimentation

K. R. DYER

An estuary can be defined as 'a semi-enclosed and coastal body of water which has a free connection with the open sea and within which sea water is measurably diluted with fresh water derived from land drainage' (Cameron & Pritchard, 1963). Thus the combination of river discharge and precipitation exceeds the loss of water due to evaporation. There is another class of estuaries where the evaporation exceeds run-off plus precipitation. These are hypersaline lagoons. They have been termed negative estuaries, in contrast to the positive estuaries defined above. The same techniques of surveying and analysis are suitable for both types of estuary, but because most estuaries are positive ones the emphasis will be placed on those.

Estuaries are very important even though their area is only a small proportion of the total world's surface. Because of their fertile waters, sheltered anchorages and the navigational access they provide to a broad hinterland, estuaries have been the main centres of man's development. However, the promotion of trade and industry has led to large-scale alteration of the natural balance within estuaries by alteration of their topography, making for easier navigation for larger ships, and large-scale pollution, as a result of industrialisation and population increases. Deforestation of the land leads to increased run-off from the land, increased flashiness of the discharge and increased sediment load in the rivers. Building and paving of large areas also produces a quick response of run-off to rainfall and increased sediment load. These effects may be controlled by the building of dams and may be reduced by the removal of river water for industrial and household use. However, control of river flow will alter the natural tendency for rivers to flush sediments out of their estuaries and consequently may aggravate shoaling problems. An example of this is the diversion of the Santee into the Cooper River, South Carolina, which altered the sediment circulation patterns and produced considerable shoaling in Charleston Harbour. To maintain adequate navigational depths now costs several million dollars a year.

Deepening of an estuary by dredging will increase the estuary volume and reclamation of intertidal areas will decrease the tidal flow, alter the mixing processes and circulation patterns, and perhaps decrease the flushing time of the estuary. With decreased flushing time the estuary cannot cope with and dispose of such large quantities of effluent. To understand and to be able to predict these effects is essential if mankind is not to do undue damage to his environment.

Prediction is based either on constructing hydraulic models or on mathematical models, into which the expected alterations can be fitted. For a fixed bed hydraulic model the topography of the estuary bed needs to be known accurately, and, for a mobile bed model, the distribution and hydraulic characteristics of the sediment as well. For both, a simple tidal curve is normally simulated at the open-sea end of the model and the roughness of the bed arbitrarily altered until the tidal curve and the flow velocities at a number of points within the estuary are faithfully reproduced. For mathematical models less precise topographic information is required, but more complete descriptions of the velocity and density fields and of the tidal elevations are needed. In all cases, however, the models are only as good as the field data upon which their characteristics are based. Because the collection of the field data is such a demanding and expensive operation it is necessary to be fully aware of the errors and drawbacks in the instruments and measuring techniques.

Origin of estuaries

Estuaries are the result of fairly recent changes in sea level and are ephemeral features since drastic alterations can be created by small changes of sea level (Russell, 1967). During the Pleistocene Ice Age much of the sea water was locked up in continental ice sheets and the sea surface stood about 100 m below its present level. In areas not covered with the ice the rivers incised their valleys to this base level and, during the ensuing Flandrian Transgression when sea level rose at about 1 m per century, these valleys became inundated. Much of the variation in form of the resulting estuaries depends on the volumes of sediment that the river or nearby coastal erosion has contributed to fill the valleys.

Where river flow and sediment discharge was high the valleys have become completely filled and even built out into deltas. Generally deltas are best developed in areas where the tidal range is small and where the currents cannot easily redistribute the sediment the rivers introduce. They occur mainly in tropical and sub-tropical areas where seasonally river discharge is very high. The distributaries or 'passes' of the delta are generally shallow and often the shallowest part is a sediment bar at their mouths. The Mississippi is an example of this type.

Where sediment discharge was less the estuaries are unfilled, though possibly they are still being filled. These drowned river valleys, or coastal plain estuaries, still retain the topographic features of river valleys having a branching, meandering outline, a triangular cross section and widening regularly towards the mouth, which is often restricted by spits. River discharge tends to be reasonably steady throughout the year and sediment discharge is generally small. These estuaries occur in areas of high tidal range where the currents have helped to keep the estuaries clear of sediment. They are typical of temperate regions such as the east coast of North America and northwest Europe, examples being the Chesapeake Bay system, the Thames and the Gironde.

In areas where glaciation was active the river valleys were overdeepened by glaciers giving fjords or inlets. A characteristic of these is the rock bar or sill at their mouths that can be as little as a few tens of metres deep. Inside, however, they can be at least 800 m deep and can extend hundreds of kilometres inland. Fjords are typical of the Canadian Pacific coast and the dimensions of many are discussed by Pickard (1961).

On low coastlines extensive lagoons are often formed with narrow connecting passages, or 'inlets', to the sea. Within the shallow lagoons the tidal currents are small, but the deeper inlets have higher currents. Again a sediment bar is generally present across the entrance. In tropical areas the lagoons can be hypersaline during the hot season. They are typical of the Southern States of the USA and of parts of Australia. The origin and classification of lagoons in Mexico is discussed by Lankford (1976).

Classification of estuaries

Estuarine processes are not simple. The river discharge is mixed with the sea water by the action of tidal motion, by wind stress on the surface and by the river discharge forcing its way towards the sea. Each of these mixing processes acts with a different time-scale of variability. The salinity difference between the river and sea water is about 35‰ creating a difference in density of about 2%. Even though this is small, it is sufficient to cause horizontal pressure gradients within the water which affect the way it flows. Density differences caused by temperature variations are comparatively smaller. Salinity is consequently a good indicator of estuarine mixing and the patterns of water circulation. Obviously there are likely to be differences in the circulation within estuaries of the same topographic type caused by differences in river discharge and tidal range. It is possible to classify estuaries according to their circulation (Pritchard, 1952; Cameron & Pritchard, 1963; Dyer, 1973). Officer (1976) and Bowden (1967) have described the mixing and circulation of estuaries.

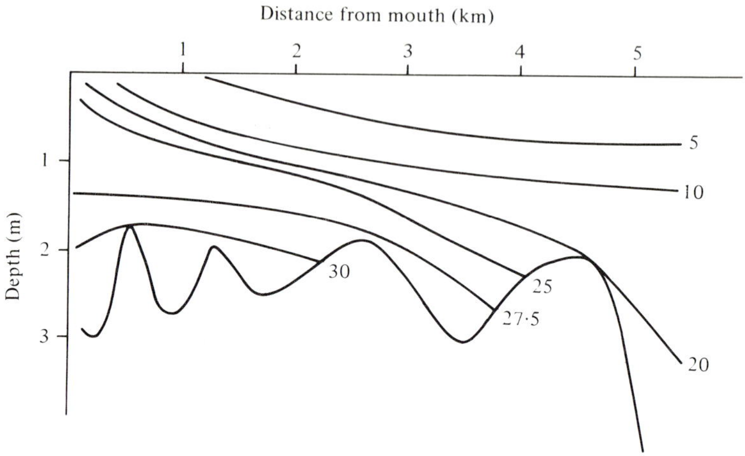

Fig. 1.1. Longitudinal distribution of mean salinity in the Vellar Estuary, a salt-wedge type estuary. Contours in‰.

Salt-wedge estuaries

The fresh water, being less dense than sea water, tends to flow outwards over the surface of the sea water which penetrates as a salt wedge along the bottom into the estuary (Fig. 1.1). This creates a vertical salinity stratification with a narrow zone of sharp salinity change, a halocline, between the two water masses, which can reach 30‰ in $\frac{1}{2}$ m. If the sea is tideless then the water in the salt wedge is almost motionless. However the surface layer flowing towards the sea, if it has a sufficiently high velocity, can create interfacial waves on the halocline. These waves break ejecting small parcels of salt water into the fresher surface layer. This process is called entrainment. In the entrainment process no fresh water is mixed downwards so that the salinity within the salt wedge is almost constant along the estuary. However the salt wedge looses salt water by mixing into the surface layer. To compensate for this upward flow, which has a magnitude of about 10^{-3} cm sec^{-1} there must be a slight landward flow within the salt wedge. Near the halocline there is a considerable velocity gradient between the strong flow towards the sea in the surface layer and the slow landward flow within the salt wedge. The position of the salt wedge will alter with the magnitude of the flow in the surface layer, i.e., with river discharge. The Mississippi River is an example of a salt-wedge estuary (Wright, 1971). When the flow in the Mississippi is low the salt wedge extends over 100 miles inland, but with high discharge the salt wedge only extends a mile or so above the river mouth. Estuaries with high river discharge in areas of restricted tidal movement can be expected to have characteristics of the salt-wedge type.

Partially mixed estuaries

When tidal movements are appreciable the whole mass of water in the estuary moves backwards and forwards with a tidal periodicity of about 12½ hours. There is considerable friction between the water and the bed of the estuary and this creates turbulence. The turbulence tends to mix the water column more effectively than entrainment, though little is known of the relationship of the turbulent exchanges of salt or momentum to the salinity and velocity gradients. Though there is no mean vertical movement of water, as there is in entrainment, there is an exchange of salt. Thus the turbulent mixing not only mixes the salt water into the fresher surface layer, but also mixes the fresher water downwards. This causes the salinity to decrease towards the head of the estuary in the lower layer and to increase progressively towards the sea in the surface layer.

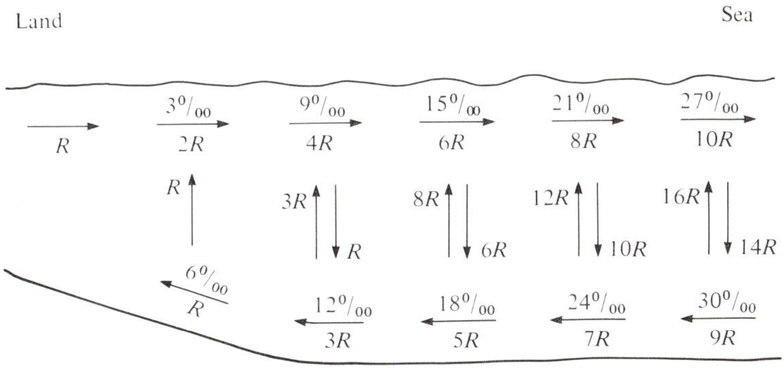

Fig. 1.2. Diagrammatic representation of mixing in a partially mixed estuary. Volume exchange in units of river flow *R*.

This process is shown diagrammatically in Fig. 1.2. In each segment there is a large turbulent exchange of water between the surface and bottom layers, in the outermost segment being 14 times the river flow. This effectively increases the salinity of the uper layer and dilutes the lower layer. Also there is an upward exchange due to entrainment which is shown as two times the river flow in all but the uppermost segment. This also increases the surface salinity and the additional volume has to be discharged seaward on the surface and provided from the landward bottom flow. Consequently at the estuary mouth a large volume of mixed water has to be discharged and the compensating inflow in the bottom layer has to be larger than in the salt-wedge estuary. However, these flows are difficult to measure because they are small mean flows on which a large tidal fluctuation is superimposed; the landward flow on the bottom can be of order 5 cm sec^{-1} with tidal oscillations of 50 cm sec^{-1}. It is necessary to average out the effect of the tidal oscillation which requires considerable precision and care in the measurements. The resulting residual or mean flow will be related to the river discharge, though the tidal response of the estuary can give additional contributions to the mean flow.

Well mixed estuaries

When the tidal range is very large there is sufficient energy available in the turbulence to break down completely the vertical salinity stratification so that the water column becomes vertically homogeneous. In this type of estuary there can be lateral variations in salinity and in velocity with a well developed horizontal circulation, or, if the lateral mixing is also intense, the estuary can become sectionally homogeneous (also called a one-dimensional estuary). In the sectionally homogeneous estuary, because there is no landward residual flow, the upstream movement of salt is produced during the tidal cycle by salty water being trapped in bays and creeks and bleeding back into the main flow during the ebb. This spreads out the salt water, but it is probably only an effective mechanisam for a small number of tidal excursions inland. The process of entrapment within shoreline irregularities has been simulated by Okubo (1973).

Fjords

Because fjords are so deep and restricted at their mouths the tidal oscillation only affects the near-surface layer to any great extent. The amount of turbulence created by the oscillation is small and entrainment is the main mixing process. Thus fjords can be considered as salt-wedge estuaries with a deep lower layer. The salinity of the bottom layer will not vary significantly from mouth to head, and the surface fresh layer is only a few tens of metres deep. When the sill is deep enough not to restrict circulation, the inflow of water occurs just below the halocline with an additional slow outflow near the bottom. When circulation is restricted the replenishment of the deeper water only happens occasionally. Renewal will occur when the coastal water at sill depth is denser than that in the fjord basin. This may be in late summer and autumn (Saanich Inlet; Anderson & Devol, 1973), in winter and spring (Skjomen; Skreslet & Loeng, 1977), it may occur regularly each flood tide (Reeburgh, Muench & Cooney, 1976) or only when the tidal currents are low (Cannon, 1975). Between inflows the density slowly decreases because of mixing, however each inflow can maintain its identity for considerable periods (Ebbesmeyer, Barnes & Langley, 1975). Gade (1973) has proposed that the rate of density decrease determines the frequency of renewal.

The descriptive classification of estuaries outlined above depends on the relative intensities of the tidal and river flows and the effect that these have on the stratification. Hass (1977) has shown how the spring–neap cycle affects the vertical salinity structure in three tributaries of Chesapeake Bay. A quantitative comparison between estuaries can be made with the diagram produced by Hansen & Rattray (1966) (Fig. 1.3). An estuary appears as a line on the diagram as the upper reaches are less well mixed than the lower sections, and because the river flow varies. This has been well illustrated for the Mersey by Bowden & Gilligan (1971).

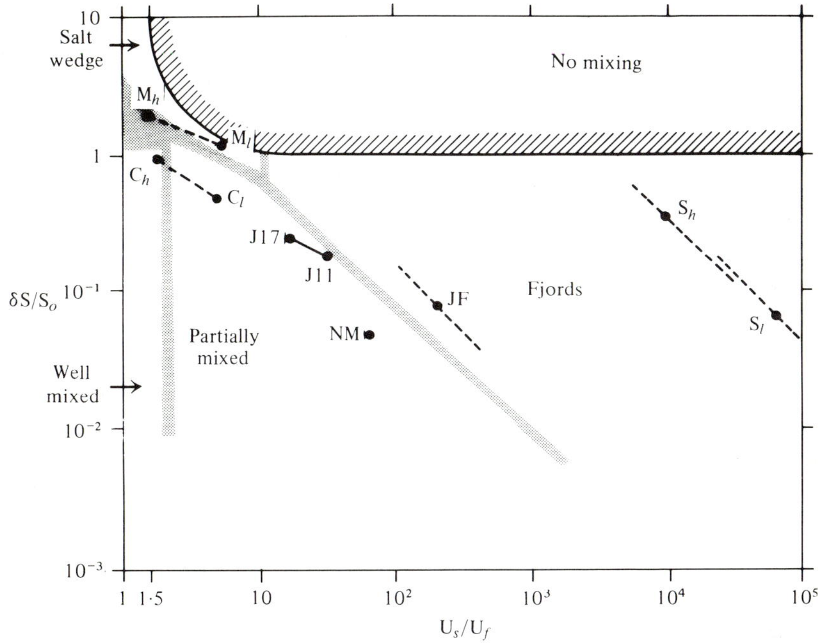

Fig. 1.3. Estuarine classification diagram (from Hansen & Rattray, 1966). δS, Difference in salinity between surface and bottom: S_o, depth mean salinity; U_s, mean surface velocity; U_f, mean flow due to river discharge; M, Mississippi River; C, Columbia River Estuary; J, James River Estuary; NM, Narrows of the Mersey Estuary; JF, Strait of Juan de Fuca; S, Silver Bay. Subscripts *h* and *l* refer to high and low river discharge. Numbers indicate miles from the mouth of the James River.

Transverse effects in estuaries

In the classification scheme outlined above there is the implicit assumption that the circulation occurs entirely in the longitudinal sense, along the estuary, and that there are no differences in the velocity or density fields across the estuary. This appears not to be strictly valid in many cases which are of importance (Dyer, 1977). The circulation will not be evenly distributed across the estuary and the water will tend to flow in a spiral fashion, creating what are known as secondary flows. These arise because of several interrelated effects, mainly the result of the topography of the estuary.

1. The cross-sectional form of the estuary is not consistent longitudinally. Thus not only does the flow speed up as the shallower water is encountered, but also the flow is deflected towards the deeper parts of the cross section.

2. Flow round bends in the estuary causes the thread of the maximum current to swing towards the outside of the bend. Bends also have changes in cross sectional form associated with them, natural deep areas being present on the outside of the bend.

3. Because the cross section of the estuary is not rectangular the vertical exchange between the saltier water in the deeper part and the fresher water is unevenly distributed.

4. Because of Coriolis Force which moves water masses towards the right in the Northern Hemisphere, there will be a tendency for the seaward-flowing surface water to keep to the right bank of the estuary.

In rivers there is a secondary circulation that gives downward flow on the outside of the bends, where there is often a scour hole, and an upward flow on the inside. Thus on a left-hand bend the secondary circulation is clockwise. In salt-wedge estuaries it appears that a similar flow pattern exists in the surface layer. However in partially mixed estuaries the balance between the various forces leads to a reversal of this situation. The seaward discharge of fresher water occurs in the shallower water and on the inside of the bends. The vertical mixing in the deeper water areas creates an upward flow and this reverses the secondary circulation giving anti-clockwise circulation on a left-hand bend. In wider estuaries, where curvature of the flow caused by bends is less important, Coriolis Force can dominate and cause the fresher water to be on the right-hand side of the estuary.

In well mixed estuaries the mixing is sufficiently intense for the residual circulation to be in the horizontal plane. In this case the seaward flowing fresher water and the landward flowing salty water are often separated by banks into 'ebb and flood' channels. Vertical variations in salinity are small and transverse effects are consequently all-important.

Consequently the transverse circulation is created by a balance between the topography, the Coriolis Force and the lateral density field. These effects have not been adequately investigated for general relationships to be clear. However the effect of transverse variations on dispersion has been treated by Fischer (1972, 1976), Okubo (1973), Prych (1970), and Smith (1978).

Effects of weather on estuaries

For many years it has been known that the classical circulation patterns are not always found in any particular estuary, but it is only recently that these effects have been seriously studied by long-term current measurements. Elliott (1976*a*) has examined the vertical variation of current velocity within the Potomac over a year. After removing the tidal velocity fluctuations he found that the dominant fluctuations in the residual currents were of a period between 2 and 5 days and with an amplitude of about 20 cm sec^{-1}. The best correlations were achieved with the wind, a downstream wind causing enhanced surface outflow and bottom inflow. The classical circulation, though the most frequently observed pattern, only occurred for 43% of the total time, with a mean duration of 2.5 days. The reverse, with inflow on the surface and outflow in the bottom, occurred 21% of the time. Total inflow, or storage, occurred for 22% of the time and the balance consisted of total discharge and three layer circulations. Most patterns presumably contained considerable lateral adjustment as well. The storage and discharge periods seem to be largely accounted for by interaction of the Potomac with Chesapeake Bay and in the result of wind effects on the coastal sea. Wind blowing southwestwards down the shelf causes a transport of water towards the shore giving an increase in sea level and an associated total inflow into the estuaries (Elliott & Wang, 1978). It appears that the classical estuarine circulation is the most efficient means of discharging the accumulated water on cessation of the wind. Similar wind effects on the circulation patterns have been found in Louisiana and Texas lagoons by Kjerfve (1975) and Smith (1977). Other studies include Elliott (1976*b*) on the Patuxent, Weisberg (1976*a*) and Weisberg & Sturges (1976) on Narragansett Bay, and Heath (1973) on a Spanish ria. Weisberg (1976*b*) has shown that to obtain a mean unaffected by the wind would require an averaging time in excess of ten days.

Wind is also an important influence in the surface layer of fjords. Pickard & Rodgers (1959) showed that up-inlet winds could reverse the surface-layer flow for a short while, presumably until the water surface had built up a slope to oppose the wind action. Outflow during this time was concentrated at the base of the surface layer. A down-inlet wind created an enhanced surface outflow and a compensating inflow between 40 and 100 m. Similar observations are reported by Farmer & Osborn (1976).

The inverted barometer effect of high atmospheric pressure depressing the water level, and vice versa, is difficult to distinguish from the other forcing mechanisms, but appears to be relatively unimportant within estuaries.

Estuarine modelling

The objective of estuarine modelling is to reproduce the major features of estuarine circulation in order that predictions can be made of, for instance, the dispersal of an effluent, or the changes created by alterations in topography. In the past much of this was done by hydraulic models, but with the advent of computers numerical models are increasingly used. Both types require validation against good quality field data.

Their use is surveyed by McDowell & O'Connor (1977) and Ward & Espey (1971).

Hydraulic models

Hydraulic models have the advantage that the natural topography can be reproduced in detail and the tidal rise and fall of the water at the open end, and the the river flow, can be varied at will. Even wind and Coriolis Force can be included. The horizontal scale is generally about a thousandth the natural size, and in order that the model reproduces the correct turbulent flow a vertical distortion is also necessary. Normally there is an order-of-magnitude difference in the horizontal and vertical scaling and thus the horizontal scale is a tenth the vertical scale.

Additionally it has been found that the details of the current velocities are not reproduced unless the salinity stratification is incorporated. In order to calibrate the model, the current velocities have to be adjusted to compare with the prototype results by altering the frictional resistance of the model. This is done empirically with metal strips or small blocks fixed to the bed. Consequently the fine structure of the flow will not match that in the prototype. Nevertheless hydraulic models produce reasonable answers to many estuarine problems.

Mathematical models

Though the mathematical equations for the estuarine circulation can be written down, they cannot be solved in their full form because there are too many unknowns. Consequently to reduce the number of unknowns various assumptions are made. These usually include some form of spatial average to reduce the three dimensional problem to two or even one dimension. The exchange ratios, about which little is known, are assumed constant or a simple variable in space and are altered so that the model fits the available data. Alternatively analytical models have been developed whereby the equations are solved for rather idealised conditions and generally applicable results obtained. For partially mixed estuaries there are studies of Rattray & Hansen (1962), Hansen & Rattray (1965) and Hansen (1967); for estuaries with three-layer flows such as Baltimore Harbour there are the solutions of Hansen

& Rattray (1972) and for fjords, Rattray (1967) and Winter (1973). These models however are limited in their ability to predict responses to alterations of the estuary.

There is a very wide range of models that have been produced for specific estuaries. Most deal with tidally averaged or steady-state conditions. Few deal with the situation within the tidal cycle; the unsteady state.

Simple models consider the mean flow to be entirely the result of river discharge and the tidal flow to be given by the tidal prism; the volume between high and low waters. Segmentation is based on simple mixing concepts and crude exchange ratios. Salinity or pollutant concentrations can then be calculated for cross sectionally averaged and vertically homogeneous conditions using the minimum of field data. These are known as tidal prism models. Examples are Ketchum (1951) and Dyer & Taylor (1973). One dimensional models are very similar, but use a finer grid system and need better data for calibration, e.g., Kent (1960) and Hamilton (1975). Two dimensional models either assume vertical homogeneity and allow lateral variations or vice versa. The simplest system to consider is a box model of two layers with fresher water overlying salty water (Pritchard, 1969). A mixed one and two dimensional model is described by Elliott (1976*c*). There are difficulties in allowing for tidally drying areas and for junctions and two dimensional models require extensive prototype data for calibration, but are more realistic. Although three dimensional models, allowing for variations along and across the estuary and with depth, have been developed (e.g., Hess, 1976), they are not extensively used because of the computational costs.

Useful insights into the relationships of estuarine characteristics with depth and river discharge (Festa & Hansen, 1976) and with mixing coefficients (Bowden & Hamilton, 1975) have been obtained by experimenting with the numerical models.

Sedimentation in estuaries

Model studies have shown that the correct sediment distribution is produced only if salinity is reproduced as well as the correct river discharge, tidal flow and topography. Consequently the estuarine circulation pattern is important in determining the sediment movement. Fine grained material will move in suspension and will follow the residual waterflow, though there may be deposition and re-erosion around about slack water. The coarser grained material will travel along the bed and will be affected most by high velocities and will move in the direction of the maximum current. Useful reviews of the literature are available in Postma (1967), Nelson (1972) and Dyer (1972).

Fine sediment

The particles smaller than about 2 μm in diameter are mainly composed of the clay minerals illite, kaolinite and montmorillonite. The particles are liable to collide and flocculate into large aggregates in sea water. The probability of this happening is a function of particle and electrolyte concentration. With adequate particle concentration flocculation of illite and kaolinite is complete above a salinity of about 4‰. For montmorillonite, however, the floc size varies over the entire salinity range (Whitehouse, Jeffrey & Debbrecht, 1960). As they can be disrupted by turbulence the size of the flocs are regulated by the intensity of the shearing in the flow. They can have diameters larger than 30 μm, but effective densities of as little as 1.1 g cm^{-3}. In sufficiently high concentrations turbulence can be damped and when the shear stress falls below about 0.8 dynes cm^{-2} the floccules will not fall as separate units, but in layers. These layers can reach concentrations as high as 300 g l^{-1} and can be observed as a distinctive 'fluid mud' on echo-sounder records (Kirby & Parker, 1974). They can flow slowly along, but once they become static the suspension slowly consolidates (Einstein & Krone, 1962). The weight of overlying sediment forces out the pore water and the floc structure collapses slowly at the bottom (Partheniades, 1965). The rearrangement of the particles gives the bed increased shear strength and resistance to re-erosion. Migniot (1968) has reviewed a large amount of data on the physical properties of fine sediments and shows that the minimum shearing stress is a function of the strength of the sediment. Thus on erosion the overlying weakly held flocs are easily entrained, and erosion continues until the shear strength of the bed balances the imposed fluid stress.

Consequently it appears that during the tidal cycle, as the current diminishes, individual or groups of flocs adhere to the bed. The mud gradually consolidates during the slack-water period and, as the current increases at the next stage of the tide, erosion may not be sufficient to remove all of the material deposited. Deposition and re-erosion must be a continuous process as there is generally more sediment in suspension than is required to complete a year's sedimentation.

A characteristic feature of well mixed and partially mixed estuaries is the turbidity maximum. This is a zone within which the suspended sediment concentrations are higher than those either in the river or further down the estuary (Inglis & Allen, 1957). It is located near the head of the salt intrusion and alters its position with changes in river discharge. The turbidity maximum is sustained by the mean estuarine circulation forming a semi-closed system. The sediment is introduced into the upper estuary by the river and the vigorous mixing exchanges the sediment into the upper layer where there is a seaward residual flow causing downstream transport. In the middle estuary the sediment settles into the lower layer in areas of less vigorous mixing and joins sediment entering from the sea on the landward residual flow. It then travels back towards the head of the estuary (Fig. 1.4). This action forms a very effective sorting mechanism and the size range of particles in the turbidity maximum is narrow (Schubel, 1969). There are also considerable areas of mud flat and tidal marshes in the area of the turbidity maximum. These also exchange considerable volumes of fine sediment with that in suspension in the estuary (Biggs, 1970). The mud flats are generally areas of deposition caused by settling over the high-water period and trapping by plants and animals. The ebb flow over the intertidal areas is concentrated into narrow creeks which migrate widely causing continual erosion. Consequently there is a continual exchange of material between one part of the marshes and another via the turbidity maximum.

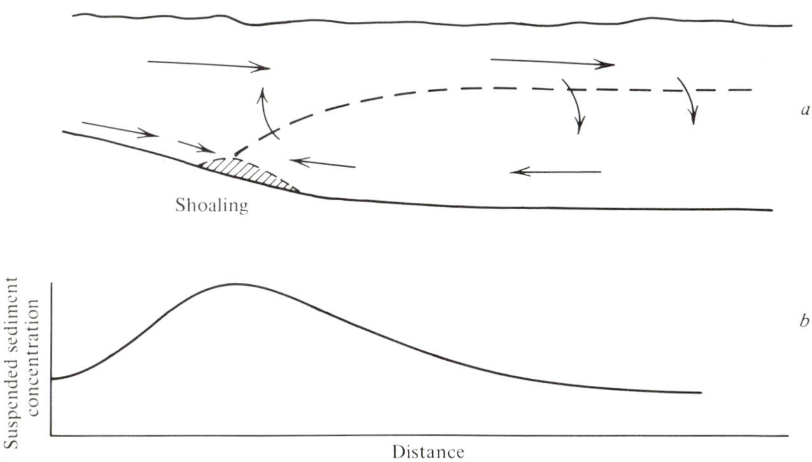

Fig. 1.4.*a*. Diagrammatic representation of circulation of suspended sediment. *b*. Resulting turbidity maximum.

Models of suspended sediment transport and the turbidity maximum have been developed by Ariathurai & Krone (1976) and by Festa & Hansen (1978) and have been reviewed by Owen (1977). There are many field descriptions including those of Nichols & Poor (1967), Allen (1973), Gallene (1974) and Buller, Green & McManus (1975).

Coarse sediments

The coarse sediments are composed of sand and gravel-size material that does not flocculate. In the river dominated reaches this bedload material will move downstream to the tip of the saline intrusion. There, as the maximum currents are equal on ebb and flood at the bottom, the material will become deposited. Further seaward the maximum landward flood current will move the bed material into a position where the velocities do not exceed the threshold for that grain size. Consequently there should be a decrease of grain size inland. However this is a rather idealised picture as the tidal currents are not simply distributed. Local constrictions will create coarse patches and widening the deposition of mud.

There can often be large lateral variations in the sediment type associated with bends and junctions. Especially in the lower part of the estuary lateral variations in velocity can be large. This can lead to sediment being transported in one direction on one side of the channel and in the opposite on the other. The flows may even become distinctly separated into ebb and flood channels, the water tending to flow up one channel and down the other, with the sediment following the same circulation system. The circulation forms an effective restriction to the penetration of coarse material into the estuary and also a good mechanism for size sorting the sediment.

Sedimentation patterns

It is apparent that the patterns of sediment movement are different in the different types of estuary and there has been considerable argument as to the importance of estuaries as sources for sedimentation on the continental shelves (Meade, 1972). In salt-wedge estuaries the river discharge of sediment is large and often markedly seasonal. Both bedload and suspended material are important. The bedload sediment will be deposited at the tip of the salt wedge but because the position of the salt wedge is so dependent on river discharge, the sediments would be spread over a long distance. The fine suspended sediments are generally carried out into the sea as a plume visible on aerial photographs (e.g., Wright & Coleman, 1971). There is no question that these rivers contribute large volumes of sediment regularly into the deeper waters.

In partially mixed estuaries the landward residual flow near the bed causes net accumulation within the estuary. Because of this there has been a fall of about 10% during the last 100 years in the volume of the Mersey estuary, despite the dredging of over 400 million cubic yards of material (Price & Kendrick, 1963). On the eastern seaboard of America there is very little sediment reaching the shelf from the estuaries. There is a predominant landward movement of sediment in most of the estuaries of that coast (Meade, 1969) and even the finest sediment that by-passes the estuaries and which can be seen as plumes is likely to accumulate further down the coast. The situation in well mixed estuaries is very similar, with accumulation within the estuary. However the large tidal ranges cause more active exchanges of the suspended sediments with the offshore waters.

All of these comments, however, are related to the normal flow conditions which, as far as the sediment transport is concerned, may not be the average. High river-discharge periods will be extremely important and will tend to flush the sediment out of the estuaries. The Delaware River carried more sediment in two days in 1955 than it carried in any other full year from 1950 to 1966. Urbanisation and other influences of man can cause the sediment discharge per unit of run-off to vary considerably (Roberts & Pierce, 1974) so that prediction of shoaling rates based on historical data may not be adequate.

REFERENCES

Allen, G. P. (1973). Etude des processus sédimentaires dans l'estuaire de la Gironde. *Mémoires de l'Institut de Géologie du Bassin d'Aquitaine, No. 5.*

Anderson, J. J. & Devol, A. H. (1973). Deep water renewal in Saanich Inlet, an intermittently anoxic basin. *Estuarine coastal mar. Sci.* **1**, 1–10.

Ariathurai, R. & Krone, R. B. (1976). Finite element model for cohesive sediment transport. *Proc. Amer. Soc. civil Eng.* **102**, 323–38.

Biggs, R. B. (1970). Sources and distribution of suspended sediment in Northern Chesapeake Bay. *Mar. Geol.* **9**, 187–201.

Bowden, K. F. (1967). Circulation and diffusion. In *Estuaries*, ed. G. H. Lauff, pp. 15–36. Amer. Assoc. Advanc. Sci. Publ. 83. Washington.

Bowden, K. F. & Gilligan, R. M. (1971). Characteristic features of estuarine circulation as represented in the Mersey Estuary. *Limnol. Oceanog.* **16**, 490–502.

Bowden, K. F. & Hamilton, P. (1975). Some experiments with a numerical model of circulation and mixing in a tidal estuary. *Estuarine coastal mar. Sci.* **3**, 281–301.

Buller, A. T., Green, C. D. & McManus, J. (1975). Dynamics and sedimentation: The Tay in comparison with other estuaries. In *Nearshore Sediment Dynamics and Sedimentation*, ed. J. Hails & A. P. Carr, pp. 201–49. John Wiley, London.

Cannon, G. A. (1975). Observations of bottom-water flushing in a fjord-like estuary. *Estuarine coastal mar. Sci.* **3**, 95–102.

Cameron, W. M. & Pritchard, D. W. (1963). Estuaries. In *The Sea*, vol. 2, ed. M. N. Hill, pp. 306–24. John Wiley and Sons, New York.

Dyer, K. R. (1972). Sedimentation in estuaries. In *The Estuarine Environment*, ed. R. S. K. Barnes & J. Green. Applied Science Publ., London.

Dyer, K. R. (1973). *Estuaries: A physical introduction.* John Wiley, London.

Dyer, K. R. (1977). Lateral circulation effects in estuaries. In *Estuaries, geophysics and the environment*, pp. 22–9. National Academy of Sciences, Washington.

Dyer, K. R. & Taylor, P. A. (1973). A simple segmented prism model of tidal mixing in well mixed estuaries. *Estuarine coastal mar. Sci.* **1**, 411–18.

Ebbesmeyer, C. C., Barnes, C. A. & Langley, C. W. (1975). Applications of an advective-diffusion equation to a water parcel observed in a fjord. *Estuarine coastal mar. Sci.* **3**, 249–68.

Einstein, H. A. & Krone, R. B. (1962). Experiments to determine modes of cohesive sediment transport in water. *J. Geophys. Res.* **67**, 1451–61.

Elliott, A. J. (1976*a*). A study of the effect of meteorological forcing on the circulation in the Potomac Estuary. Chesapeake Bay Inst., Johns Hopkins Univ., Special Report 56.

Elliott, A. J. (1976*b*). Response of the Patuxent Estuary to a winter storm. *Chesapeake Sci.* **17**, 212–16.

Elliott, A. J. (1976*c*). A mixed-dimension kinematic estuarine model. *Chesapeake Sci.* **17**, 135–40.

Elliott, A. J. & Wang, D-P. (1978). The effect of meteorological forcing on Chesapeake Bay: the coupling between an estuarine system and its adjacent waters, pp. 127–45. In. *Hydrodynamics of estuaries and fjords*, ed. J. C. J. Nihoul. Elsevier Oceanography Series, No. 23.

Farmer, D. M. & Osborn, T. R. (1976). The influence of wind on the surface layer of a stratified inlet: Part 1. Observations. *J. phys. Oceanog.* **6**, 931–40.

Festa, J. F. & Hansen, D. V. (1976). A two-dimensional numerical model of estuarine circulation: the effects of altering depth and river discharge. *Estuarine coastal mar. Sci.* **4**, 309–23.

Festa, J. F. & Hansen, D. V. (1978). Turbidity maxima in partially mixed estuaries: A two-dimensional numerical model. *Estuarine coastal mar. Sci.* **7**, 347–59.

Fischer, H. B. (1972). Mass transport mechanisms in partially stratified estuaries. *J. Fluid Mech.* **53**, 672–87.

Fischer, H. B. (1976). Mixing and dispersion in estuaries. *Ann. Rev. fluid Mech.* **8**, 108–33.

Gade, H. G. (1973). Deep water exchanges in a sill fjord: a stochastic process. *Phys. Oceanog.* **3**, 213–19.

Gallene, B. (1974). Study of fine material in suspension in the estuary of the Loire and its dynamic grading. *Estuarine coastal mar. Sci.* **2**, 261–72.

Hamilton, P. (1975). A numerical model of the vertical circulation of tidal estuaries and its application to the Rotterdam Waterway. *Geophys. J. Roy. astron. Soc.* **40**, 1–21.

Hansen, D. V. (1967). Salt balance and circulation in partially mixed estuaries. In *Estuaries*, ed. G. H. Lauff, pp. 45–51. Amer. Assoc. Advanc. Sci. Publ. 83. Washington.

Hansen, D. V. & Rattray, M. Jr. (1965). Gravitational circulation in estuaries. *J. mar. Res.* **23**, 104–22.

Hansen, D. V. & Rattray, M. Jr. (1966). New dimensions in estuarine classification. *Limnol. Oceanog.* **11**, 319–26.

Hansen, D. V. & Rattray, M. Jr. (1972). Estuarine mixing induced by diffusion. *J. mar. Res.* **30**, 281–94.

Hass, L. W. (1977). The effect of the spring–neap tidal cycle on the vertical salinity structure of the James, York and Rappahannock Rivers, Virginia, U.S.A. *Estuarine coastal mar. Sci.* **5**, 485–96

Heath, R. A. (1973). Flushing of coastal embayments by changes in atmospheric conditions. *Limnol. Oceanog.* **18**, 849–62.

Hess, K. W. (1976). A three-dimensional numerical model of the estuary circulation and salinity in Narragansett Bay. *Estuarine coastal mar. Sci.* **4**, 325–38.

Inglis, Sir C. C. & Allen, F. H. (1957). The regimen of the Thames estuary as affected by currents, salinities and river flow. *Proc. Inst. civil Eng.* **7**, 827–68.

Kent, R. E. (1960). Diffusion in a sectionally homogeneous estuary. *Proc. Amer. Soc. civil Eng.* **86**, SA2, 15–47.

Ketchum, B. H. (1951). The exchanges of fresh and salt water in tidal estuaries. *J. mar. Res.* **10**, 18–38.

Kirby, R. & Parker, W. R. (1974). Seabed density measurements related to echo sounder records. *Dock and Harbour Authority*, **54**, 423–4.

Kjerfve, B. J. (1975). Tide and fair weather wind effects in a bar-built Louisiana Estuary. In *Estuarine Research*, vol. 2, ed. L. E. Cronin, pp. 47–62. Academic Press, New York.

Lankford, R. R. (1976). Coastal lagoons of Mexico – their origin and classification. In *Estuarine Processes*, Vol. 2, ed. M. Wiley, pp. 182–215. Academic Press, New York.

McDowell, D. M. & O'Connor, B. A. (1977). *Hydraulic behaviour of estuaries*. Macmillan, London.

Meade, R. H. (1969). Landward transport of bottom sediments in estuaries of the Atlantic Coastal Plain. *J. Sed. Petrology*, **39**, 222–34.

Meade, R. H. (1972). Sources and sinks of suspended matter on continental shelves. In *Shelf Sediment Transport: Process and pattern*, ed. D. J. P. Swift, D. B. Duane & O. H. Pilkey, pp. 249–62. Dowden, Hutchinson & Ross, Stroudsberg.

Migniot, C. (1968). Study of the physical properties of various forms of very fine sediments and their behaviour under hydrodynamic action. *La Houille Blanche*, **23**, 591–620.

Nelson, B. W. (ed.) (1972). Environmental framework of coastal plain estuaries. *Geological Soc. America, Memoir 133*.

Nichols, M. M. & Poor, G. (1967). Sediment transport in a coastal plain estuary. *Proc. Amer. Soc. civil Eng* **93**, WW.4, 83–95.

Officer, C. B. (1976). *Physical oceanography of estuaries (and associated coastal waters)*. John Wiley, New York.

Okubo, A. (1973). Effect of shoreline irregularities on streamwise dispersion in estuaries and other embayments. *Neth. Jour. Sea. Res.* **6**, 213–24.

Owen, M. W. (1977). Problems in the modelling of transport erosion and deposition of cohesive sediments. In *The Sea*, ed. E. D. Goldberg, I. N. McCave, J. J. O'Brien & J. H. Steele, pp. 515–37. John Wiley, New York.

Partheniades, E. (1965). Erosion and deposition of cohesive soils. *Proc. Amer. Soc. civil Eng.* **91**, HY1, 105–39.

Pickard, G. L. (1961). Oceanographic features of inlets in the British Columbia mainland coast. *Jour. Fish. Res. Bd. Canada*, **18**, 907–99.

Pickard, G. L. & Rodgers, K. (1959). Current measurements in Knight Inlet, British Columbia. *Jour. Fish. Res. Bd. Canada*, **16**, 635–78.

Postma, H. (1967). Sediment transport and sedimentation in the estuarine environment. In *Estuaries*, ed. G. H. Lauff, pp. 158–79. Amer. Assoc. Advanc. Sci. Publ. 83, Washington.

Price, W. A. & Kendrick, M. P. (1963). Field and model investigation into the reasons for siltation in the Mersey estuary. *Proc. Inst. civil Eng.* **24**, 473–517.

Pritchard, D. W. (1952). Estuarine hydrography. *Adv. Geophysics*, **1**, 243–80.

Pritchard, D. W. (1954). A study of the salt balance in a coastal plain estuary. *J. mar. Res.* **13**, 133–44.

Pritchard, D. W. (1969). Dispersion and flushing of pollutants in estuaries. *Proc. Amer. Soc. civil Eng.* **95**, HY1, 115–24.

Prych, E. A. (1970). *Effects of density differences on lateral mixing in open-channel flows.* Rep. KH-R-21. W. M. Keck Laboratory, California Institute of Technology.

Rattray, M. Jr. (1967). Some aspects of the dynamics of circulations in fjords. In *Estuaries*, ed. G. H. Lauff, pp. 52–62. Amer. Assoc. Advanc. Sci. Publ. 83. Washington.

Rattray, M. Jr. & Hansen, D. V. (1962). A similarity solution for circulation in an estuary. *J. mar. Res.* **20**, 121–33.

Reeburgh, W. S., Muench, R. D. & Cooney, R. T. (1976). Oceanographic conditions during 1973 in Russell Fjord, Alaska. *Estuarine coastal mar. Sci.* **4**, 129–46.

Roberts, W. P. & Pierce, J. W. (1974). Sediment yield in the Pataxent River (Md.) undergoing urbanization, 1968–1969. *Sediment. Geology.* **12**, 179–97.

Russell, R. J. (1967). Origins of estuaries. In *Estuaries*, ed. G. H. Lauff, pp. 93–9. Amer. Assoc. Advanc. Sci. Publ. 83. Washington.

Schubel, J. R. (1969). Size distributions of suspended particles of the Chesapeake Bay turbidity maximum. *Neth. J. Sea. Res.* **4**, 283–309.

Skreslet, S. & Loeng, H. (1977). Deep water renewal and associated processes in Skjomen, a fjord in North Norway. *Estuarine coastal mar. Sci.* **5**, 383–98.

Smith, N. P. (1977). Meteorological and tidal exchanges between Corpus Christi Bay, Texas and the northwestern Gulf of Mexico. *Estuarine coastal mar. Sci.* **5**, 511–20.

Smith, R. (1978). Coriolis, curvature and buoyancy effects on dispersion in a narrow channel. In *Hydrodynamics of estuaries and fjords*, ed. J. C. J. Nihoul, pp. 217–32. Elsevier Oceanography Series No. 23.

Ward, G. H. & Espey, W. H. (ed.) (1971). Estuarine modelling: An assessment. No. 16070 DZV02/71. Environmental Protection Agency, Washington.

Weisberg, R. H. (1976*a*). The nontidal flow in the Providence River of Narragansett Bay: a stochastic approach to estuarine circulation. *J. Phys. Oceanog.* **6**, 721–34.

Weisberg, R. H. (1976*b*). A note on estuarine mean flow estimation. *J. mar. Res.* **34**, 387–94.

Weisberg, R. H. & Sturges, W. (1976). Velocity obvservations in the West Passage of Narragansett Bay: a partially mixed estuary. *J. phys. Oceanog.* **6**, 345–54.

Whitehouse, U. G., Jeffrey, L. M. & Debbrecht, J. D. (1960). Differential settling tendencies of clay minerals in saline waters. *Proc. 7th Conf. Clays, Clay Min.* 1–79.

Winter, D. F. (1973). A similarity solution for steady-state gravitational circulation in fjords. *Estuarine coastal mar. Sci.* **1**, 387–400.

Wright, L. D. (1971). Hydrography of South Pass, Mississippi River. *Proc. Amer. Soc. Civil Engrs.* **97**, WW3, 491–504.

Wright, L. D. & Coleman, J. M. (1971). Effluent expansion and interfacial mixing in the presence of a salt wedge, Mississippi river delta. *J. geophys. Res.* **76**, 8649–61.

2

Tidal measurement

N. C. GLEN

Tidal variations in the water level in estuaries are generally greater than those in the open sea. There can also be large variations within any one estuary due to funnelling, slowing down of the tidal wave, and frictional dissipation. The funnelling can lead to an increase of range in the outer parts of the estuary and then the frictional dissipation to a decrease near the head. The water slopes, which are important in echo sounding, are extremely variable and without proper tidal observations any estimate is liable to error. In general terms the levels at high water can be expected to be much more nearly horizontal than those at low water.

Tidal measurements are required for the following purposes:—

1. For the construction of navigational charts, echo-sounding surveys must be converted to a datum. Initially observations should be carried out to establish the level of the lowest astronomical tide, at which the datum is generally placed. Once this has been done tide measurements will need to be taken during the surveys.

2. To provide predictions for navigational purposes continuous records are necessary. Many estuaries show large changes in water level with atmospheric pressure and wind. For vessels operating with minimal underkeel clearance actual, rather than predicted levels may be required.

3. For tidal research a period of at least a month is required. This will give the astronomical constants for prediction and may provide empirical weather correction factors. Many tide gauges are operated to give surge and flood warnings.

Tidal observations

Generally at least two tide gauges are required, one at the mouth of the estuary and the other at the head. As many sites as possible should be observed simultaneously, preferably all those being used for the area of the estuary. This is particularly important in an estuary where the levels are liable to be markedly affected by changes in the river flow. Corrections can be made to allow for the changes of river flow but these invariably reduce the accuracy of the final results.

The period of observations required for most purposes is one month. Fewer data are of very little use, more can be of value but should be in units of a month at a time. The first step is therefore to carry out a reconnaissance of the estuary to choose sites which are suitable for observations and which will give a good coverage of the area. Points to note are the ease of accessibility as well as ease of installation of the gauge.

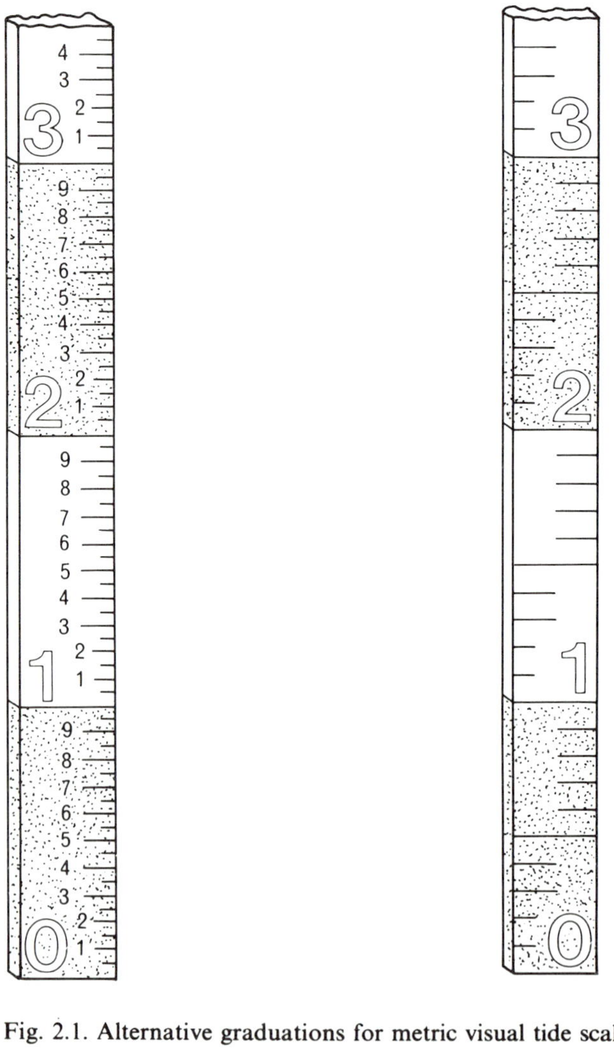

Fig. 2.1. Alternative graduations for metric visual tide scales.

Visual tide scale

The simplest form of tide gauge is a pole marked in metres and decimetres, usually known as a Visual Tide Scale or tide pole (Fig. 2.1.). Care must be taken to ensure that when the pole is installed it is truly vertical. This is comparatively simple when a pier or wharf is available, though it may be necessary to offset the top of the pole to ensure verticality. When this is not possible it is essential to measure and record the inclination, so that the necessary corrections can be made to all the observations. The best way of avoiding errors is to construct a scale drawing. This drawing should also show the datum of the land levelling system, such as Ordnance Datum (Newlyn) in Britain, and any other information which will enable the comparison of observations between sites to be made without error (Fig. 2.2). After analysis it is also possible to insert the tidal levels, such as Mean High Water Springs etc., in their correct relative positions. A visual tide scale is necessary even if it is intended to install an automatic tide gauge, as it will always be needed for the calibration of the gauge and connection to the land levelling system.

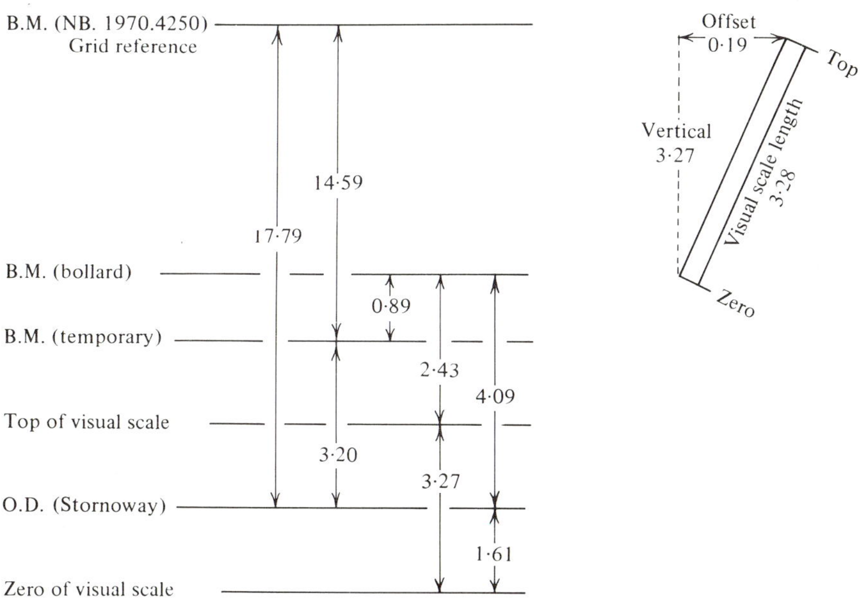

Fig. 2.2. Record of installation and levelling data for a visual tide scale.

It is important that the Visual Tide Scale is erected so that it is still possible to obtain readings even at the lowest low water. In an estuary this is particularly difficult as many piers and jetties dry out at low water. Nevertheless, care must be taken to ensure that impounding does not occur and that the level recorded truly represents the actual level of the estuary. This can only be determined by comparing readings along the estuary and by use of Low Water Plots (p. 35). Frequently it is necessary to install off-lying poles to enable true readings to be obtained at low water. At most sites in a survey of the River Severn it was found impossible to cover the full range of the tide, over 10 m, with a single pole. One pole could be erected on the lowest low-water line to cover the whole range but, at high water, it would not be possible to get nearer than several hundred metres except by boat. Readings from the shore, even using binoculars, would be too inaccurate. Much the simplest, and most accurate method, is to erect a number of shorter poles in steps down the shore. These should be arranged so that two poles can be read simultaneously as a check as the observers advance and retreat up the shore. Care must then be taken over the level connections between the poles and it is convenient if these are included on the drawings referred to above. The outer poles must obviously be rigidly secured or guyed, and not interfere with shipping. The least number of observers required to obtain continuous readings day and night for a month is three.

Automatic tide gauges

The problem of finding sufficient staff for lengthy tidal observations leads to the use of automatic tide gauges, although the cost of the instruments and their installation is often considerable. Provided the gauge is working well and carefully set up it is usually sufficient to visit the site once each day to carry out the essential checks, wind the clock and also change the recorder paper. The less frequently the gauge is visited, the less reliable the records become. Complete reliance should never be placed on results from a gauge which is not properly maintained.

Tide gauges are generally available in two types, float and pressure gauges. There are also some gauges that measure water level by the changing electrical resistances between two wires or by the changing capacitance of a single wire, but these are sophisticated scientific instruments used for extremely precise tidal research.

Float gauges

These consist of a vertical tube or stilling well, open to the sea through a small orifice near the bottom and in which a float sits on the water surface. A wire joins the float to a recording mechanism (Fig. 2.3), and it is kept taut by a counterbalance weight, or sometimes a spring. By a system of gears a pen is moved across the recorder paper as the height of tide changes. The paper is secured round a drum which is rotated by clockwork and the height of tide is recorded against time. Some instruments are now available which record

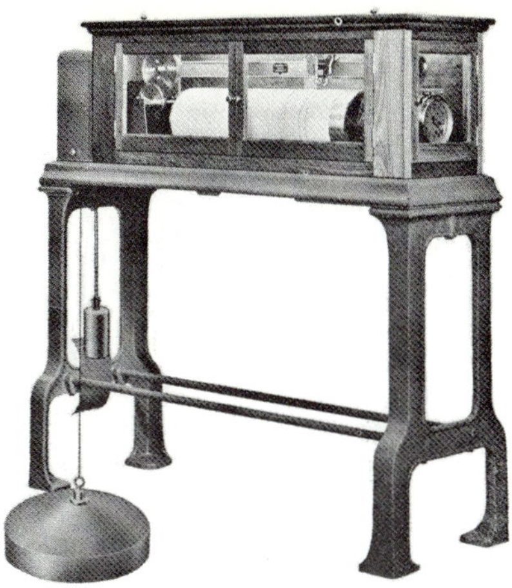

Fig. 2.3. Admiralty horizontal drum-type tidal recorder (without stilling well). Photograph courtesy of R. W. Munro, Ltd.

on punched paper or magnetic tape at regularly timed intervals, however these usually require electricity supplies which makes them less convenient to use. An assessment of the design and optimum dimensions of the float tube has been made by Halliwell & Perry (1969). A convenient size for the orifice is a diameter one tenth that of the well. This ensures a reasonably free flow of water into the well so that there is no difference in height between the water inside and outside. At the same time it smooths out the larger surface waves in the estuary and the record is reasonably free from rapid movement. Care must be taken to see that the orifice is kept clear of mud and weed; permanent gauges have the orifice regularly cleared by divers to ensure accurate records.

The operation of a tide gauge should be frequently checked by a test known as the Van de Casteele Test (Lennon, 1968). This consists of lowering a measuring line down the well with an electric circuit indicating when the line touches the water. The measurement of the height of the water surface below a fixed point at the top of the well is added to the height of tide shown by the recorder. If the gauge is working correctly this sum should always be the same at any stage of the tide. Thus a plot of these results against the height of the tide, taken on both the rising and falling tides, will show very clearly if there is any backlash in the mechanism, or any other recorder error. It does not, however, ensure that the gauge is recording the true level of the estuary. This can only be ensured by careful comparison of readings from the gauge with those obtained on the Visual Tide Staff. These are needed because not only may the orifice become blocked by weed, but other sources of error may also be present.

One error which is difficult to detect is caused by strong tidal streams in the vicinity of the well orifice. The thrust of the moving water may increase or decrease the pressure at the orifice, and so cause the levels of the water inside and outside the well to differ. Provided that the tidal streams are less than one knot ($\frac{1}{2}$ m sec^{-1}) it can usually be assumed that these differences are not significant (Halliwell & Perry, 1969).

Another cause of error arises from changes in the density of water in the estuary. The water inside the well has a density which depends on that of the water entering the orifice. Because of the estuarine stratification and lags in the adjustment of water temperature in the well, this can have a different pattern to that of the water outside, particularly at high water. This will create a difference in level between the water inside and outside the well. An estimate of the importance of this error can be obtained from a series of density profiles in the estuary and comparison with measurements of salinity and temperature in the well (Lennon, 1966).

Temporary float gauges can be installed but the work required to set up the well and ensure that it is working properly is very considerable (Noye, 1974). The site also has to be chosen very carefully; usually a jetty with vertical sides is necessary and the foot of the jetty must not dry out at low water. These conditions are not often found so that most float tide gauges will be at permanent sites. Even in these cases careful checking of the readings is essential if accurate data are to be obtained.

Checking Float Gauges. The method to be described applies only to analogue gauges, digitally recording gauges require a special system of checks. Thorough recording of the various checks for later correction of the tidal levels is essential. Fig. 2.4 shows a useful daily check sheet for which the recommended procedure is:

(a) Daily (preferably about the same time):

(i) Read the visual tide staff, allowing for wave motion, and note the time using a reliable clock;

(ii) Read the tidal height and time as shown on the recorder graph sheet, as well as the actual time if different from that above;

(iii) Do not attempt to adjust the gauge for any discrepancy but ensure that these are fully noted on the check sheets.

(b) Weekly (preferably Monday morning):

(i) Carry out the usual daily procedure as described above, entering the details on the last line of the Check Sheet;

(ii) Lift the pen clear of the graph sheet and remove the drum. Now cut the paper along the 22.30 or 23.30 h time line. Most drums are provided with a longitudinal cutting groove which must be used when cutting the graph sheet to avoid scoring the working surface of the drum;

SUMMARY OF CHECKS ON AUTOMATIC TIDE GAUGE

Week commencing 24-1-77

Place BRIDLINGTON

Time kept (G.M.T. or B.S.T. or Zone) G.M.T.

Date and time of starting record 25-1-77 1200

Date and time of removing record 31-1-77 0910

Date	Correct Time	Time on Chart	Height of tide on tide scale	Height of tide on Chart	Scend on tide scale	WIND		Remarks (including times of sudden changes in wind force or direction)	Initials of Checker
						Direction	Force		
25-1-77	1400	1402	1.26	1.26	0.03	S	3-5	WIND INCREASING STEADILY	[illegible]
26-1-77	1120	1125	3.62	3.80	0.05	E	7	WIND INCREASING AND BACKING	[illegible]
27-1-77	1730	1737	2.18	2.25	0.10	NW	7-8		E.B.B.
28/1/77	0800	0810	3.00	3.08	0.20	N	7-8	CLOCK 10 MINUTES FAST TIME CORRECTED AT 0810	[illegible]
29/1/77	0900	0902	2.84	2.90	0.01	NE	2-3		[illegible]
30-1-77	1600	1604	2.94	3.13	0.15	SW	4-5		E.B.B.
31-1-77	0900	0905	1.69	1.74	0.10	SW	3-4		[illegible]

[illegible]
Officer in charge of gauge

Fig. 2.4. Example of daily check sheet for automatic tide gauge.

(iii) Note the date and time on the old graph sheet;

(iv) Fully wind the recorder clock and check the ink in the pen. When the nib of a pen becomes blocked and requires cleaning, this should be done with methylated spirits. It is essential that the correct type of pen for the gauge is used. If the pen has become worn or the wrong type is used, it may become loose in the holder, tremors will be produced in the ink line, and it may be assumed, quite incorrectly, that the disturbances have been produced by the tidal motion. The tidal range is also likely to be reduced;

(v) Enter on the new recorder graph sheet the date and approximate time at which it is being put on;

(vi) Before sticking down the new graph sheet ensure that the bottom end beds down on the flange of the drum all round, and that the height graduations are in line all along the join. Make sure that the join is the right way, i.e. the pen will slide over it and will not stick;

(vii) Replace the drum (if removed to take off the old graph sheet) and set the pen to the correct time by rotating the drum. The final movement of the drum must be anti-clockwise in order to take out the backlash;

(viii) At least 5 min after setting the pen on the graph sheet (to allow for all the backlash to have been taken out) compare the time shown on the graph with the correct time and if necessary re-set the graph;

(ix) Read the Visual Tide Staff. If this does not now agree with the tide gauge, first check to ensure that the pen has not been accidentally displaced. If so, a new calibration will be necessary. Although the Visual Tide Staff provides a rough check of the gauge on a daily basis it cannot be read with sufficient accuracy to enable the gauge to be accurately re-set;

(x) As a final check, compare the time and height readings on the old graph with those on the new graph to ensure that there is no gross error (i.e. 1 h). Repeat the usual daily check.

Faults in Float Automatic Tide Gauges. Some common faults and likely causes are indicated in Table 2.1.

Table 2.1

Symptom		Cause
1. The trace shows a prolonged 'flat' at high water	(1)	Undue friction in the mechanism
	(2)	Counterweight too light
	(3)	Water in float
	(4)	Insufficient travel for counterweight
	(5)	Float well not vertical, causing friction between float and well

2. The trace shows a prolonged 'flat' at low water	(1)	As (1) to (5) above
	(2)	Silting at base of float well, causing float to ground
	(3)	Base of well not low enough causing float to ground
	(4)	Tide gauge well impounded
3. Steps in trace (always at or near same height)	(1)	Small obstruction in float well
	(2)	Counterweight fouling obstruction
	(3)	Wires or bands fouled or kinked
4. Steps in trace (random)	(1)	Friction in mechanism, probably due to lack of oil in moving parts
5. Sudden vertical jump in height	(1)	Due to kink or fouled wire freeing itself
	(2)	Penholder loose
6. Trace remains at same height for prolonged period (except at high or low water)	(1)	Inlet blocked by floating obstruction
7. Sudden horizontal jump in time	(1)	Penholder loose
	(2)	Drum has been rotated by hand, or knocked
8. Pen failing to mark	(1)	Ink supply needs replenishing
	(2)	Pen needs cleaning
	(3)	Insufficient pressure on paper
	(4)	Spring, where fitted to penholder, broken
	(5)	Drum not vertical (where failure to mark is either at top or bottom only)
9. (a) Rapid vertical oscillations in trace, which will be thick and irregular	(1)	Inlet too large
	(2)	Float well leaking, fractured or removed
(b) Ditto – but only above a certain height	(1)	Water entering well over the top
	(2)	Gap between sections of well
10. Recorded levels persistently lag behind true levels	(1)	Inlet too small
	(2)	Inlet partially obstructed
11. Checks show cyclic time error over every 24 h period	(1)	Drum not circular
	(2)	Drum not vertical, or similar manufacturing fault

Pressure Tide Gauges

These gauges are easier to install than float gauges but in most cases are not so accurate and have their own calibration difficulties. There are four varieties of pressure gauge:

Trapped Air Gauges. These consist of a diaphragm box held securely in position under water with a flexible membrane separating the air from the water outside. The upper part of the box is connected by a flexible nylon pipe to a pressure capsule, a Bourdon Tube, or a mercury manometer which operates some form of recorder. The pressure caused by the head of water above the diaphragm is transmitted by the air in the pipe to the recorder. The effect of friction on the air in the pipe reduces the the effect of waves. It also affects the accuracy of the records so that it is necessary to keep the length of the pipe down to between 30 and 60 m. Longer pipes have been used with success, but usually only in areas where the range of the tide is comparatively small. The final accuracy obtained is largely dependent on the method used for measuring the pressure. The mercury manometer is undoubtedly the most accurate, giving results to about 0.01 m of water, but it is rather more difficult to install. The simpler type of gauge uses either a Bourdon Tube or a pressure unit, similar to that used in an aneroid barometer. These gauges are generally more robust but tend to be somewhat less accurate though, with care, results to about 0.02 or 0.03 m can be obtained.

With any pressure gauge of this type it is absolutely essential that the connections between the tube and both the recorder and the diaphragm box are made completely airtight. The diaphragm box must be rigidly secured at a level which is always under water and the same remarks regarding the impounding of the gauge apply as for any other kind of gauge. It is also necessary to ensure that the box is not installed in an area where strong tidal streams may actually impinge on the box. If this occurs there will be a dynamic pressure acting on the diaphragm which may cause considerable errors in the recorded level. In the River Severn, with tidal streams up to about 4 knots or more, it was found impossible to use a pressure gauge for this reason.

Open-ended gauges. The open-ended type of gauge is very similar but instead of using a flexible membrane between the air and the water, the water is allowed to find its own level in the diaphragm box. There is very little to choose between these two types for ease of installation but the open-ended type requires a rather larger size of box below water. This is needed to act as a buffer for the wave action and also to provide sufficient air to ensure that water never enters the nylon tube, otherwise the gauge will fail completely. The only additional difficulty encountered is due to the fact that, as the air in the pipe is compressed, the air/water interface moves up the box and so there is a slight movement of the zero of the gauge. This movement depends on the height of tide and is taken into account during the calibration. Also the air in the tube gradually becomes saturated with water so that there is a gradual rise in the zero of the gauge. Provided that the observations do not last longer than about one month this source of error can usually be disregarded (Pugh & Alcock, 1976).

Bubbler gauges. The Bubbler gauge has an open-ended pipe with air, or some other convenient gas such as nitrogen, supplied under pressure at the upper end of the system. The flow of the gas is carefully controlled so that only about one bubble a second enters the system. Thus the pressure due to the head of water in the estuary and that from the gas cylinder are equalised at the open end of the tube, which must, of course, be rigidly fixed. The flow of gas must be just sufficient to keep the water out of the tube but not so great as to introduce a back pressure caused by friction in the pipe. The major advantage of this type of gauge is that the pipe can usually be longer than with the Trapped Air type. However, there is one disadvantage which is peculiar to the Bubbler gauge. When a large wave passes over the open end of the tube, there is an increase of pressure followed very quickly by a decrease as the trough passes. The air comes out of the tube very quickly when the pressure outside is reduced, but only builds up slowly as the pressure increases, due to the controlled rate of air supply. Before the pressure has been able to build up again the next trough is passing over the gauge. Thus the level recorded by the gauge is somewhat lower than the actual level of the sea if there were no waves present. The amount of the reduction is dependent on both the amplitude and period of the waves. Correction for this error is difficult to apply. One cure is to increase the rate of supply of the gas, but the suitable rate is very difficult to judge, as it will vary with weather conditions and will also reduce the life of the gas bottles. A more effective solution is to ensure that this type of gauge is only used in positions where there is complete shelter from all large waves. Under these conditions it is a very useful gauge and has the advantage that it is not necessary to secure a large box under water, only the open end of a small nylon tube.

Electric strain gauges. A pressure unit containing electric strain gauges is connected to a recorder, that can either be on the end of an electric cable or in a watertight self-recording unit on the sea-bed. These gauges are very sensitive, but also need compensation for variations in atmospheric pressure before the recorded pressures can be converted to heights of water. However, they have the advantage that the cable can be very long indeed or can operate offshore when used as an independent unit. The calibrating and checking of these instruments is usually more complicated and they are normally only used in experienced oceanographic laboratories.

Clocks

In addition to the possible errors in height recording which may occur in any of the gauges, it must also be remembered that it is equally important to ensure that the time is accurately measured and recorded. On all the gauges mentioned the time is recorded by a clockwork or an electrical clock rotating a drum, to which the recorder paper is attached, or a disc of recording paper. In order to ensure time accuracy it is essential to run all the clocks for a period before the survey and carefully rate them.

Scales

The accuracy of the records will depend not only on the efficiency of the various mechanisms, but also on the scales used in the analogue records. The scale of heights which is most suitable is between 5 and 10 cm to represent 1 m. For the time scale about 3 cm to the hour is commonly used. When the paper is secured to a drum, or when a disc of paper is used, the clock should rotate the record once every 24 h. The tide moves on by about 45 min each day so that it is possible to use the same chart for several days. Charts are normally changed weekly but, if the range of the tide is small, or there are large wave effects, it may be necessary to change the recorder paper more frequently to ensure that the records are legible.

Calibration

It is essential that any tide gauge is calibrated to ensure that the heights are correctly recorded, or any required corrections are known. Soon after installing the gauge and the visual tide staff, a series of comparisons between the two should be recorded over a complete tidal cycle. If possible this cycle should be near spring tides to cover the full range of the tide gauge. These are plotted as shown in Fig. 2.5 and a straight line drawn through the points. Any good recorder should have a straight line calibration, the slope of which is dependent on both the instrument in use and the site at which it is installed. Should the recorder be changed or moved to a new site, a new calibration immediately becomes necessary. As the observations progress it is possible to add the daily checks to this diagram. If these checks are made at about the same time each day, they will be well spaced out over the calibration line. They thus check the initial calibration and also indicate when any change of level in the zero takes place, e.g. by loss of air pressure or movement of the diaphragm box.

A new calibration will need to be carried out after any repair.

Sometimes it is not possible to carry out a calibration over the full range of the tide. This is often the case when the only possible site for the tide pole dries out at extreme spring tides. Provided that some low water checks can be made, and obviously the more of these that are possible the better, it can usually be assumed that there is a linear relation between the actual level of the sea and the visual tide scale readings. Extrapolation is then possible to give a calibration over the full scale required. The accuracy of this partial

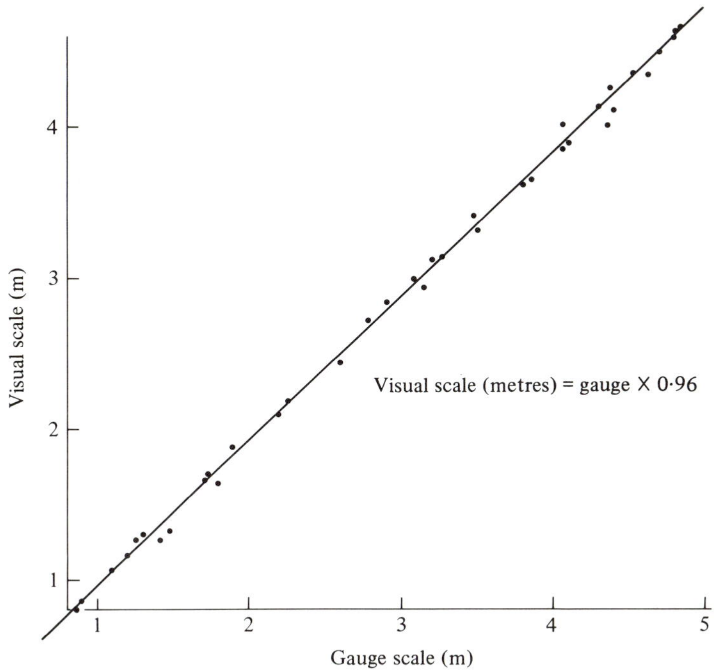

Fig. 2.5. Calibration of automatic tide gauge against visual scale, Bridlington, Jan–Feb 1977.

calibration will depend on the range over which the observations have been completed and should only be used when a full calibration is not possible. It is also a wise precaution to check the response of the pressure recorder under laboratory conditions before starting the survey to ensure that the assumption of linearity is justified.

Datum levels

In order to compare the results of tidal observations at different sites in an estuary it is necessary to have a basic horizontal plane of reference which extends throughout the area. This is normally the level of the Land Levelling System of the country; on the mainland of the United Kingdom it is Ordnance Datum (Newlyn). The level is usually established by a long period of tidal measurements at one particular site from which a level such as the Mean Sea Level can be obtained. In the case of OD(N) it is the mean level of the sea obtained from the average of hourly heights of the tide between 1916 and 1921 at Newlyn. This level is accurately connected to some very stable Bench Marks in the vicinity of the tide gauge which are then accepted as defining the fundamental level. It must be noted that MSL itself is not necessarily a stable level over long periods of time (Lisitzin, 1974). In the southern part of England MSL is rising at a rate of about 2 mm per year, though this rate is probably not constant in time or over the whole area. Thus in 1977 MSL at Newlyn is about 0.1 m above OD(N).

A large number of Bench Marks throughout the mainland of UK are connected by accurate surveying to the fundamental level. It has also been possible to connect some of the nearest islands to the system but further islands, such as the Outer Hebrides and Shetlands, have their own systems. Each visual tide staff should be connected by local levelling to one, or preferably two, of these Bench Marks and thus to the national levelling system. This enables a diagram to be drawn of an elevation through the estuary (Fig. 2.6). A horizontal line is drawn to represent the level of OD(N) and the position of each visual tide staff can be plotted together with its associated tidal levels.

This diagram is an essential part of the survey as it shows the tidal levels at points betweeen the actual sites of observation. This is nearly always necessary as it is seldom possible to observe at all the points where data are required. It also shows clearly the slopes of the various tidal levels. It should be noted that in the example Fig. 2.6, mean low water spring tides (MLWS) becomes a level slightly *above* that at neap tides (MLWN) at the head of the estuary. This is not uncommon and appears to be due to the relationship between the area of mudflats covered by the tide in the upper reaches of the estuary and the surface friction experienced by the tide during the ebb part of the cycle.

In areas where no land levelling system exists the only alternative is to assume that MSL as obtained at each of the tide gauges is a horizontal plane. This is not generally satisfactory, particularly in the upper reaches of an estuary where the assumption is almost certainly not valid. In these areas a local levelling system between the sites of tidal observations may have to be set up if the survey is to provide reliable results. An alternative technique is described by Pugh & Alcock (1976).

The other important level which is usually involved in this work is Chart Datum (CD). This is the level on navigational charts to which soundings are related. It is often necessary to use depth information from these charts in the course of the survey and therefore this datum should be included in the diagram of tidal levels (Fig. 2.6). Chart Datum is commonly defined as a 'level so low that the tide will not frequently fall below it'. In those areas for which the UK Hydrographer of the Navy is the surveying authority, and also in some others, this is interpreted as a level below which the tides will not be predicted to fall. This level is also known as Lowest Astronomical Tide and is defined in detail in Admiralty Tide Tables (ATT).

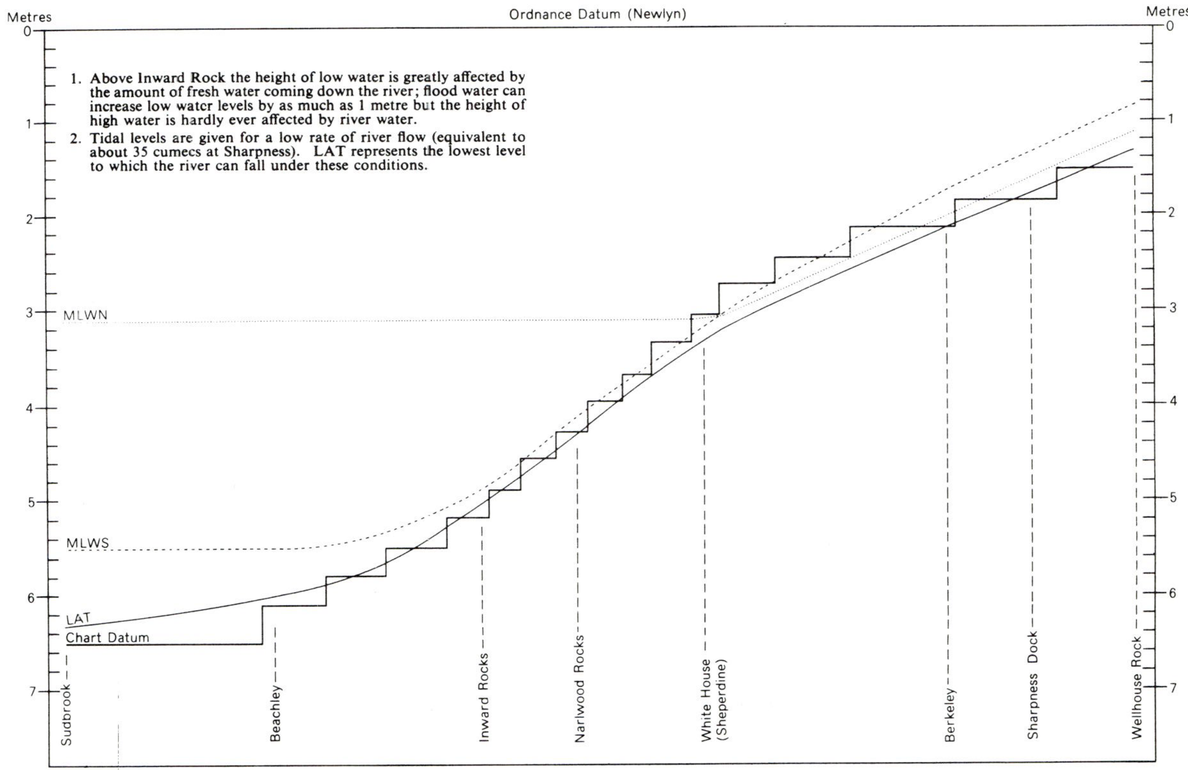

Fig. 2.6. Diagram showing chart datum and low-water levels between Sudbrook, and Sharpness, Severn Estuary (referred to Ordnance Datum, Newlyn). Produced from portions of BA Chart no. 1166 with the sanction of the Controller HM Stationery Office and of the Hydrographer of the Navy.

From the definition it will be seen that the connection between MSL or OD(N) and CD will vary from place to place in accordance with the range of the tide. To assist in establishing CD in those areas where a Land Levelling System exists, the connection between the two is usually published. The connection between CD and OD(N) in UK is published both on the largest scale Admiralty Chart and also in Admiralty Tide Tables for all those places where it is known. For places where CD is not known, it can either be found by interpolation on the diagram or from the analysis of the tidal observations. The levels obtained from the analysis of well-controlled tidal observations are always to be preferred (Admiralty Tidal Handbook No. 2).

Tidal analysis

At any site the tidal curves can be analysed for the various lunar and solar constituents. Once these are known, predictions of the water levels can be made. The measurements are affected by the weather and these variations must be removed during the analysis so that the constants obtained are only those relating to the tide. There are two basic methods of analysis, Harmonic and Non-Harmonic. The results obtainable from a Harmonic Analysis are severely limited by the length of data available. For a fully comprehensive analysis of a site in an estuary, observations over a complete year are necessary. This is due to the complex nature of the tidal curve in these areas and the large number of higher harmonics required to describe it adequately. A smaller number of constants, which may be sufficient in some cases where the distortion of the curve is less, can be obtained from one month's data. A well-tried method for this analysis is published in Admiralty Tidal Handbook No. 1.

Non-Harmonic Analysis can be used in areas with distorted curves and where data for only a short period are available. It involves comparing the observations at the various sites with simultaneous observations at a Standard Port where long period data are available and a detailed analysis has already been done. The comparisons are done graphically and provide relative levels and time differences which describe the tidal movements in the estuary to a reasonably high standard of accuracy. The work can be done by computer but the graphical method is preferred as, in addition to giving accurate results, an inspection of the plots obtained gives a very good idea of the accuracy of the data and frequently enables the researcher to detect errors which would otherwise have passed unnoticed. It also shows when the tide pole or gauge has become impounded.

By plotting observations of the heights of High Water observed at the Standard Port against those for the same events at each site in the estuary

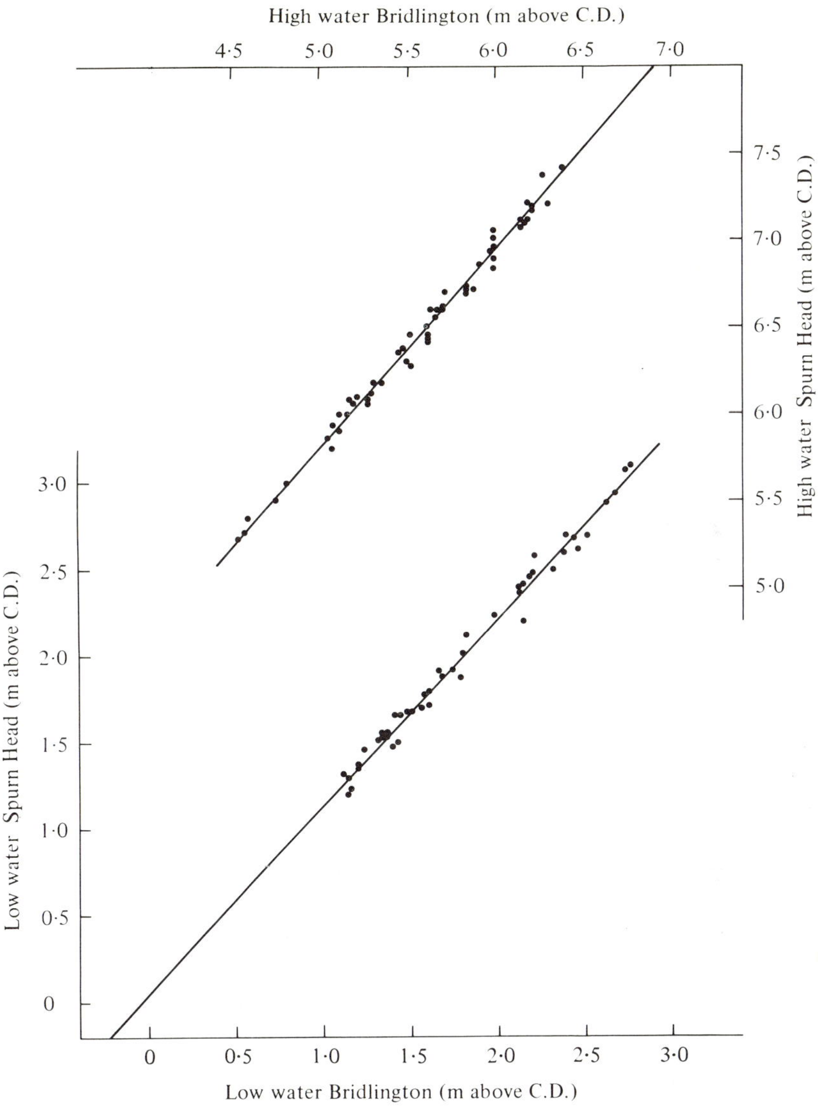

Fig. 2.7. Low-water and high-water plot of Standard Port (Bridlington) against estuarine tide gauge (Spurn Head).

a simple correlation can be made, as illustrated in Fig. 2.7. Similar plots are made for Low Water Heights, High Water Times and Low Water Times. The plots will show a certain amount of scatter, depending to a large extent on the suitability of the Standard Port. The scatter is largely caused by the differences in the meteorological effects at the two places being compared.

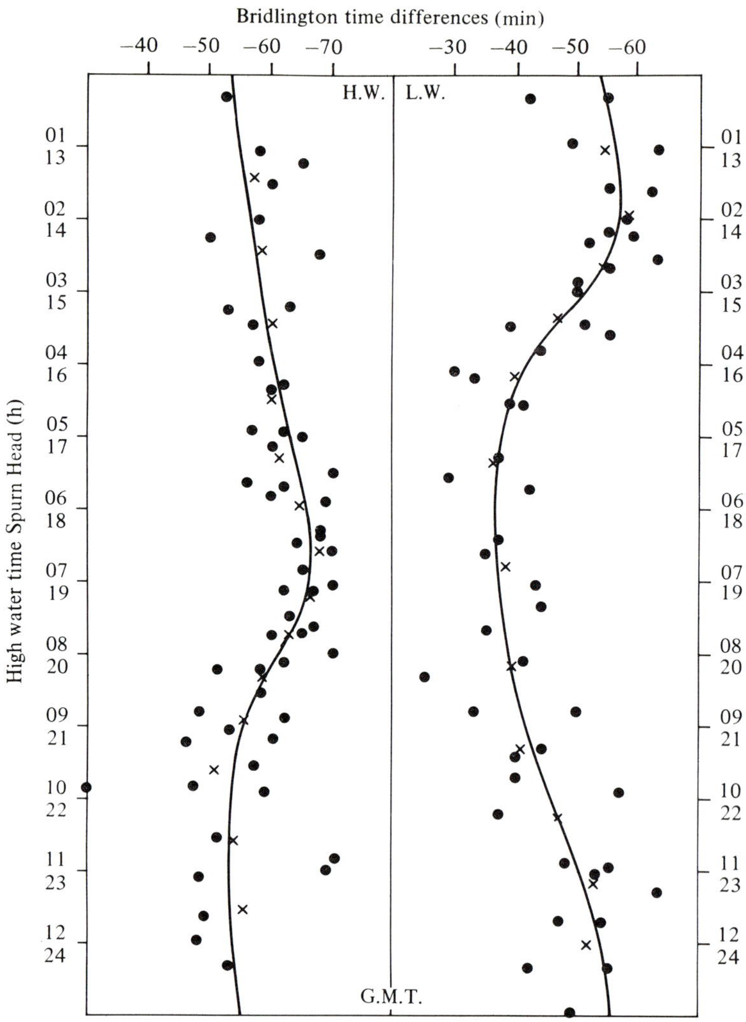

Fig. 2.8. Plot of low-water and high-water times at Standard Port (Bridlington) against estuarine tide gauge (Spurn Head). Dots are observations, crosses are block averaged results.

It is also usually larger in the time plots than the height plots. If the scatter is very large another Standard Port should be considered. The correlation can be calculated using the principle of 'Least Squares'. The time differences are usually plotted against the time of High Water at the Standard Port and a curve of an approximately sinusoidal type drawn. (Fig. 2.8).

The height plots will also indicate sites where the low water levels are not actual tidal levels but dependent on the level of river flow. From the gauging data held by the Water Authority it is often possible to obtain a correlation between the volume of river flow and the low water levels recorded during the survey. Thus the observed low waters can be corrected to the level expected at a certain rate of river flow (see Fig. 2.6). The standard used is normally a dry weather rate with an indication given of the amount of rise to be expected in rainy conditions. Similar effects are seldom found in the High Water levels in an estuary but changes in these levels are often found in the rivers themselves.

Tidal waves in estuaries

Every body of water has a natural period of oscillation that is dependent on the dimensions of the basin. If one considers a rectangular basin, the normal form will consist of a nodal point in the centre and two anti-nodes, one at each end. If the dimensions of a gulf or ocean are such that this period is similar to one of those of the tide-raising forces, two or one cycles per day, there is every possibility that resonance will occur. The tide-raising forces themselves are very small and thus large tidal movements, which are often found in nature, will only arise when there is resonance.

If one considers this basin in more detail one finds that, at the two ends, where the vertical motion is greatest, the horizontal motion or tidal streams are nil; while at the centre, where the vertical motion is least, the tidal streams are greatest. In the real situation these movements are taking place on the surface of the rotating earth. This introduces the Coriolis Force which affects the movements in much the same way as the winds around an area of low barometric pressure are affected. Thus the nodal line across a rectangle becomes a point in the centre, usually called an Amphidromic Point, where the tidal movement is least. The wave of high water rotates around this point, with the maximum vertical tidal movements at the outer limits of the area. The time of maximum velocity occurs at the same time as mean water level with slack water at high and low water. Not many estuaries are sufficiently large for the effect of Coriolis Force to be important, but some of the very large Gulfs have suitable dimensions. Cases of interest are Hudson's Bay, the Gulf of Carpentaria and particularly the Persian Gulf. For thc last of these an Atlas is published showing details of the movements in both the diurnal and semi-diurnal modes (Admiralty Chart 5081).

The other kind of wave, which is more common in estuaries, is the progressive wave. This is the result of tides generated in the ocean which then propagate into and up the estuary, and finally into the river where the energy is dissipated and the effects cease to be noticeable. One of the important points with this type of wave is that the tidal streams are greatest near the time of high or low water and least near half-tide.

In most estuaries, however, the actual movements are a combination of

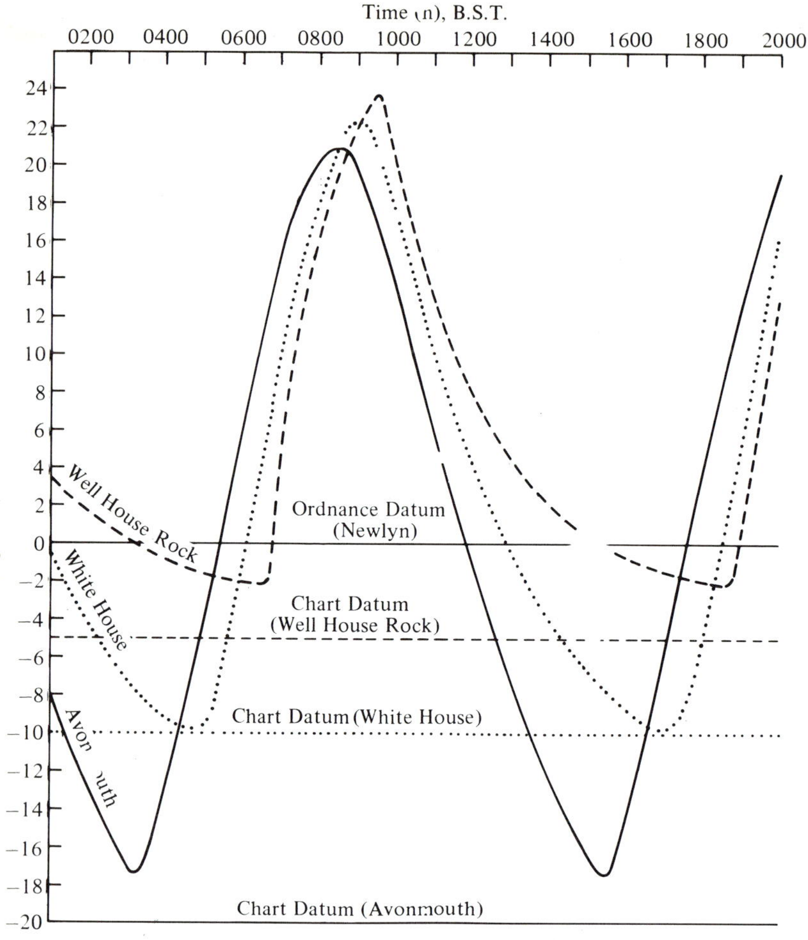

Fig. 2.9. Tidal curves for three positions on the River Severn on 20 April 1962 showing progressive distortion of the tide within the estuary.

standing and progrcssive waves. The slack waters occur near the times of high and low water while there is a steady progression of high water up the estuary. These effects can be used to characterise estuaries (Ippen, 1966).

In most estuaries the depth of water is not very great, particularly at low water. This causes considerable friction between the water and the sea-bed, which is most noticeable during the falling tide. This can be easily seen on tide-gauge records where the slope of the rising tide will be much steeper than that of the falling tide (Fig. 2.9). Also the duration of the rise of the tide will be much shorter than that of the fall.

The shallow-water effects described above cause the crest of the tidal wave to travel up the estuary much faster than the trough. In some estuaries the effect of friction, combined with the slope of the sea-bed, reach critical values at which bores are formed (Tricker, 1964). In the critical cases the onset of the crest of the tide takes the form of a breaking wave which can be clearly seen advancing up the estuary. This wave can reach a height of 2 m in the River Severn, though larger ones can be found in other rivers.

In addition to the obvious feature of the breaking wave, a very rapid reversal of the tidal stream takes place as the wave passes. Thus a 3 knot outgoing tidal stream is replaced by one of the same speed in the opposite direction during the few seconds taken by the passing bore. This effect combined with the very rapid rise in water level can be very dangerous to the unwary observer collecting data in the estuary.

REFERENCES

Admiralty Tidal Handbook No 1. (1959). The Admiralty Semi-Graphic Method of Tidal Analysis. NP 122(1).

Admiralty Tidal Handbook No 2. (1975). Datums for Hydrographic Surveys.

Admiralty Chart 5081. (1976). Persian Gulf. Tidal Atlas.

Halliwell, A. B. & Perry, J. G. (1969). Hydrodynamic studies of tide gauges. *J. hydraulics Res.* **7**, 485–517.

Ippen, A. T. (1966). *Estuary and coastline hydrodynamics.* McGraw-Hill, New York.

Lennon, G. W. (1966). The behaviour of stilling wells in the presence of periodic density variations. *Institute of Oceanographic Sciences (Bidston) Internal Report No. 7.*

Lennon, G. W. (1968). The evaluation of tide gauge performance through the Van de Casteele test. *Cahiers Oceanographique*, **XX**, 867–77.

Lisitzin, E. (1974). *Sea Level Changes.* Elsevier Oceanography Series No. 8.

Noye, B. J. (1974). Tide Well Systems. *J. mar. Res.* **32**, 129–94.

Pugh, D. T. & Alcock, G. A. (1976). Gradients of the mean and instantaneous sea surface level in a shallow water region. *Int. hydrographic Rev.* **LIII**, No. 2, 29–39.

Tricker, R. A. R. (1964). *Bores, breakers, waves and wakes.* Mills & Boon Ltd.

GENERAL READING

Admiralty Manual of Tides (1941). NP 120. Ministry of Defence, Hydrographic Dept., Taunton.

Admiralty Tide Tables; Vol. 1, European Waters. NP 201 (published annually).

Admiralty Tide Tables; Vol. 2, Atlantic and Indian Oceans. NP 202 (published annually).

Admiralty Tide Tables; Vol. 3, Pacific Ocean and Adjacent Seas. NP 203 (published annually).

Admiralty Manual of Hydrographic Surveying 1969. **2**, Chap. 2. Tides and Tidal Streams. NP 134b(2).

Admiralty Method of Tidal Prediction. (1975). NP 159.

Admiralty Chart 5057. (1974). Dungeness to Hoek van Holland. Co-Tidal Chart.

Admiralty Chart 5058. (1974). British Isles and Adjacent Waters. Co-Tidal Chart.

Admiralty Chart 5059. (1971). Southern North Sea. Co-Tidal Chart.

Admiralty Chart 5082. (1973). Malacca Strait. Co-Tidal Chart.

Admiralty Tidal Stream Atlas. (1970). Thames Estuary (with Co-Tidal Chart) NP 249.

Admiralty Marine Science Publication No. 7. (1963). Mean Sea Level in the British Isles NP 650(7).

Canadian Hydrographic Service. (1964). OTTBORO Tide Gauge. *Int. hydrographic Rev.* **XLI**, No. 2, 165–74.

Cross, R. H. (1968). Tide Gauge Frequency Response. *Proc. Amer. Soc. Civil Eng.* **94**. WW3, 317–30.

Glen, N. C. (1976). Co-Tidal Charts. *The Geographic Journal*, **142**, 554–5.

Glen, N. C. (1971). Tide Gauges. Conference of Commonwealth Survey Officers. Report of Proceedings Part II, *Paper F6*.

Glen, N. C. (1977). The Admiralty Method of Tidal Prediction NP 159. *Int. hydrographic Rev.* **LIV**, No. 1, 73–85.

Gordon, D. L. (1967). Rationalisation of Chart Datum. Conference of Commonwealth Survey Officers. Report of Proceedings, Chap. 4, Paper C 1.

Gordon, D. L. (1960). The Accuracy of Automatic Tide Gauges. *Bulletin Geodesique*, **55**, 59–63.

Lennon, G. W. (1967). A critical examination of the Conventional Tide Gauge. International Hydrographic Organisation. Proceedings of the Symposium on Tides, Paris.

Pugh, D. T. (1972). The Physics of Pneumatic Tide Gauges. *Int. hydrographic Rev.* **XLIX**, No. 2, pp. 71–97.

Sigl, R. (1970). Report of the Symposium on Coastal Geodesy. Institute for Astronomical and Physical Geodesy. Technical University, Munich.

Symposium on Tidal Instrumentation and Predictions of Tides. (1965). International Association for the Physical Sciences of the Oceans. Special Publication No. 27.

Symposium on Tide Recording, Proceedings. (1976). The Hydrographic Society, Special Publication No. 4.

3

Hydrographic surveying

D. J. HOOPER

> Passenger on Mississippi steamer: 'I suppose you are the man who knows where all the shoals are'.
> Pilot: 'No ma'am, I'm the man who knows where they ain't'.
> Mark Twain

Hydrographic surveying in estuaries tends to be conducted along the lines of the above quotation. Port authorities, pilots and shipmasters are mainly interested in safety of navigation through approach channels and for this reason, combined with the limited time available over high water, the banks and shoals are but seldom surveyed with the same thoroughness.

The marine scientist may well therefore require to conduct some preliminary sea-bed mapping as a prelude to his research, and will undoubtedly need to know with some accuracy the position or positions where his data is collected. This chapter is designed to help him achieve this by providing an elementary introduction to hydrographic surveying.

Position

There is a wide variety of methods for fixing position offshore. These vary in complexity and cost, from the simple optical sextant fix to complex electronic positioning systems associated with computers and data loggers. The more common methods likely to be used in estuaries are summarised in Table 3.1.

Optical methods

Double horizontal sextant angle resection

The two angles subtended by three fixed marks, either on land or projecting from the water, are simultaneously measured with hand-held sextants by two observers standing close together on the survey vessel. The position derived from these two angles may be plotted on a chart by use of station pointers. This instrument has two adjustable arms on either side of a fixed centre arm which can be set at the two angular values observed, and the instrument is

Table 3.1. *A summary of the more common methods for positioning in estuaries*

1. *Optical methods*
   - (i) Double horizontal sextant resection
   - (ii) Single horizontal sextant angle and transit line
   - (iii) Theodolite intersection
2. *Electronic methods*
   - A. Two megahertz (nominal) systems
     - (i) Decca HiFix and HiFix 6
     - (ii) Decca SeaFix and MiniFix
     - (iii) Sercel Toran
     - (iv) Raydist
   - B. Microwave systems
     - (i) Tellurometer MRB 201
     - (ii) Decca Trisponder D202
     - (iii) Cubic Autotape
     - (iv) Motorola Mini Ranger III
   - C. Range and bearing system
     - (i) Artemis
3. *Combined system*
   - (i) Theodolite bearing and microwave distance

then manoeuvred on a plotting sheet until the three arms pass through the plotted positions of the fixed marks. The centre of the instrument then coincides with the fix position of the observer.

Alternatively, a previously prepared lattice chart of arcs of circles of constant subtended angles may be used as an underlay to the plotting sheet (Fig. 3.1).

Station pointers have the disadvantage that they are easily damaged and are, arguably, the greatest source of error in this method of position fixing. They are, however, more versatile than lattice charts which, although avoiding this source of error and being quicker in use, restrict the observer to the shore points previously chosen.

The following rules should be borne in mind when using this method for position fixing:

(i) The observing vessel should not lie on, or close to, the circle that passes through the three points observed to, as, in this case, the two position circles in Fig. 3.1 would coincide. Consequently the centre one of the three fixed marks should be the closest to the observing vessel or, if this is not practicable, it should be much further away than the other two. This is to avoid the 'danger circle' situation.

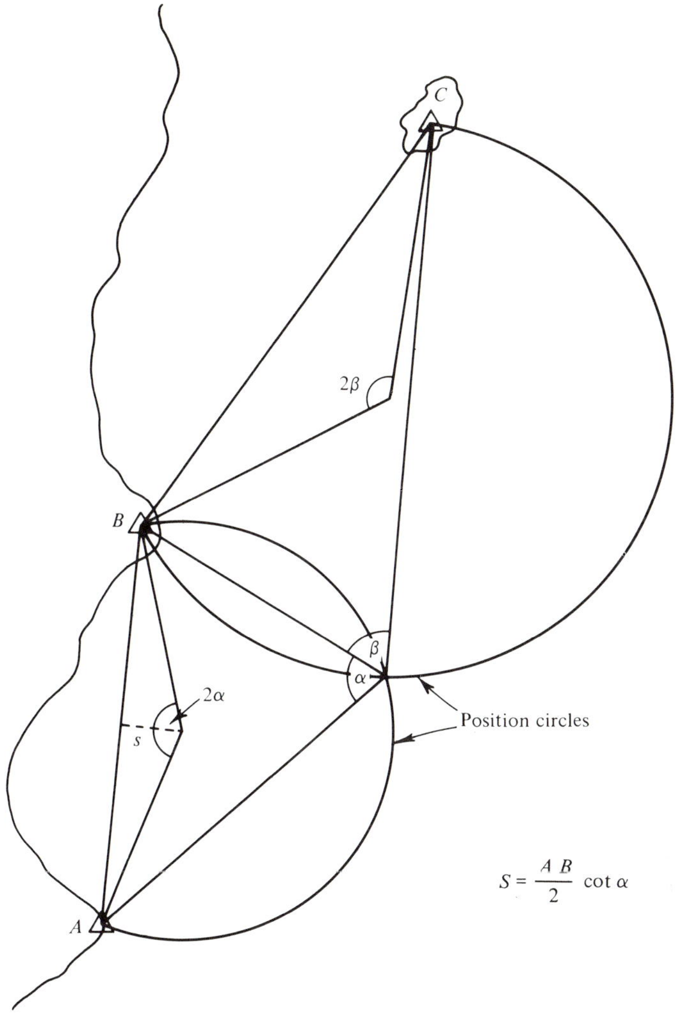

Fig. 3.1. Construction of arcs of constant subtended angle. *A*, *B*, *C*, three fixed marks. $S = \frac{AB}{2} \cot \alpha$.

(ii) The fixed marks should lie at or close to sea level.

(iii) Observed angles should lie between 30° and 140°.

(iv) A good fix will be obtained if the observing vessel lies inside the triangle formed by the three points observed to, provided that the subtended angles are not less than 60°.

Table 3.2. *Characteristics of optical positioning methods*

Method	Resolution	Accuracy	Remarks
1. Double horizontal sextant angle resection	1 min of arc, each angle	(at best) ±1 m per km of range	(1) Requires two observers (2) Both must observe simultaneously and stand close together
2. Single horizontal sextant angle and transit	1 min of arc, (i) for angle, (ii) depends on ratio of distance between transit marks and distance to observer	(at best) ±1 m per km of range	Transit distance should be not less than 25% of observing distance
3. Theodolite intersection	1 s of arc	±1 m	Good communication essential

N.B. *All optical positioning systems require line of sight and good visibility*

Single horizontal sextant angle and transit line

In this method a single horizontal sextant angle is observed across a fixed transit line between two natural or artificial points (Fig. 3.2). This method limits the observer to positions along the transit line but is useful where repeated observations are to be made along the same line. It has the advantage that it can be operated by a single observer on board the survey vessel, provided that the helmsman can keep the vessel on the survey line. It is necessary therefore that the transit should be reasonably 'sensitive', i.e. that a small deviation from the line will be immediately evident from the separation of the two marks. This requires MN to be a reasonably long distance relative to NP.

Ideally, the two sextant marks should be equally spaced on either side of the transit line, so that the arcs of constant subtended angle cut it as nearly at right angles as possible.

Position fixes may be plotted either by setting the single angle on station pointers and plotting across the transit line or as a pre-drawn lattice of circles of constant subtended angles and transit line.

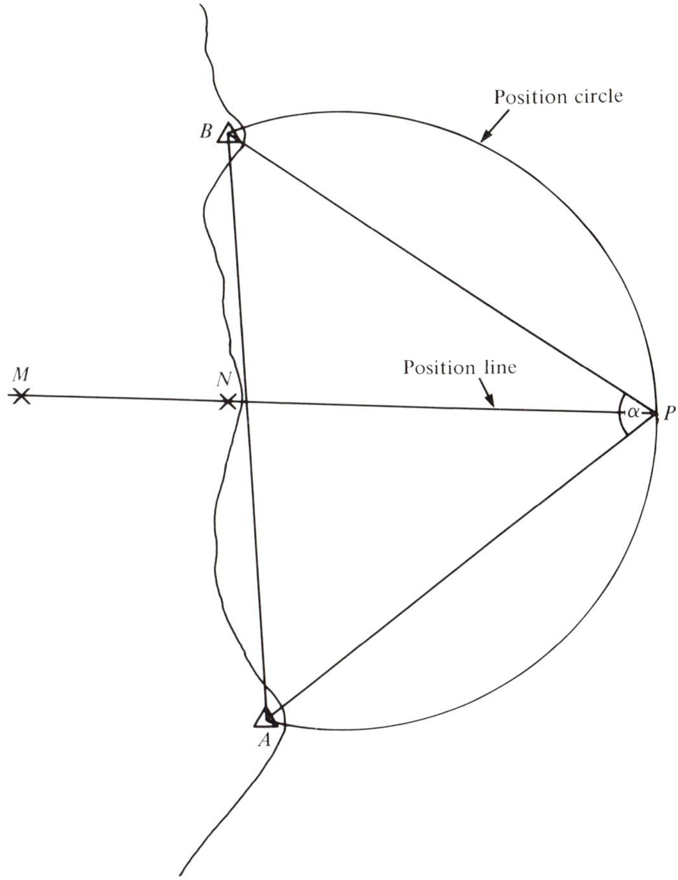

Fig. 3.2. Single horizontal angle and transit line. *A*, *B*, sextant marks; *M*, *N*, transit marks; α, subtended angle.

Theodolite intersection from shore

Two theodolite bearings are observed simultaneously from two known points on shore, between fixed reference objects and a common point on the vessel (Fig. 3.3). This method requires good communications between vessel and shore observers to ensure simultaneous measurement of the two angles together with the depth measurement. Positions may be plotted graphically on a pre-drawn chart of bearing lines radiating from the two theodolite positions. It is often convenient if A and B are at the ends of a measured base line and the angles BAP and ABP are observed.

A base-line length producing a large angle of cut between the theodolite range is desirable. If this angle is small then the base-line length needs to be known very precisely.

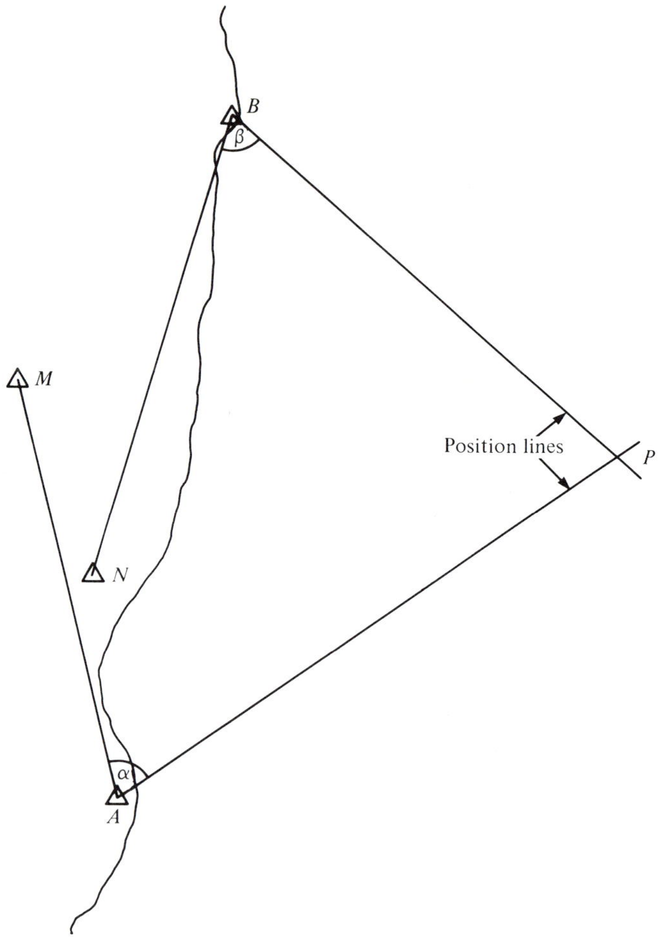

Fig. 3.3. Theodolite intersections from shore. *A*, *B*, theodolite stations; *M*, *N*, reference objects; α, β, observed angles.

The advent of programmable desk-top calculators has made it possible to circumvent the use of lattice charts for plotting optically determined positions. Survey co-ordinates (bearings and/or angles) may be converted to map co-ordinates (distance east and north from a single origin) for plotting on a millimetric grid. The limitations of pre-drawn lattices and the errors arising from the use of station pointers may thus be avoided, although care is necessary to avoid being unwittingly in the 'danger circle' situation.

Electronic positioning methods

Electronic position fixing systems cover the whole of the frequency spectrum from visible light down through infra-red and microwave frequencies to Very Low Frequency systems at 10 kHz or so. The systems commonly in use in estuaries fall generally into the 2 MHz and the 3000 to 9500 MHz bands and this section will be limited to an outline description of these.

Two megahertz systems

These systems are summarised in Table 3.3. They all employ the process of phase comparison to measure distance. The difference in phase between radio signals transmitted synchronously from two separate transmitters is measured and displayed as a fraction of the wavelength of the signals. The incoming signals will be in phase (i.e. at zero phase difference) at every half wavelength from the shore transmitters along the base line joining them. Thus, these systems do not measure directly the number of points of zero phase difference. Ambiguity may be resolved by superimposing a coarser frequency over the basic signal, but this is seldom done on the grounds of cost and the difficulty of obtaining clearance for the use of the additional frequencies.

Consequently, phase comparison systems in the 2 MHz band are designed to integrate the whole lane crossings as the receiver moves through the area of coverage. Thus the correct whole-lane value is correctly displayed anywhere in the area of coverage provided that the correct value is set in at a known reference point. Whole-lane count may be lost as a result of electro-magnetic interference, or of power loss on board ship. Whole-lane values should therefore be checked as often as possible, and certainly no less frequently than at the beginning and end of each survey period.

Two megahertz systems are employed generally either in the hyperbolic mode with, typically, a Master and two Slave stations on shore, or in the two-range mode where the Master station is situated on the vessel (Fig. 3.4).

Microwave positioning systems

The most commonly used microwave positioning systems are summarised in Table 3.4 for convenience. The Artemis system is included in this section. Artemis is a range and bearing system, providing a display of these parameters from a single Remote station. The other four systems all measure and display ranges from two Remote stations.

The Tellurometer MRB 201 and the Cubic Autotape systems make their measurements by comparing the phase of the incoming signal from the Remote station with that of the transmitted signal at the shipborne Master. A series of coarser frequencies are superimposed on the basic frequency to overcome range ambiguity.

Table 3.3. *Characteristics of typical 2 MHz positioning systems*

System	Number of frequencies (kHz)	Resolution (m)	Type of transmission	Nominal maximum range (km)	Configuration	Maximum no. of users	Resolution of ambiguity	Power requirements
HiFix	1 1600–3000	0.5 to 0.9	Interrupted CW	90–360	2 Range Hyperbolic	1 Multi	Yes, but all equipment must be duplicated	144w at 24vdc (shore) 96w at 24vdc (ship) (hyperbolic mode)
SeaFix and MiniFix	1 1600–3000	0.6 to 0.9	Interrupted CW	40–100	2 Range or 3 Range Hyperbolic	1 Multi	No	15w at 24vdc (shore) 48w at 24vdc (ship) (hyperbolic mode)
HiFix 6	1 1600–3000	0.5 to 0.9	Interrupted CW	300	2 Range Hyperbolic (Compound configurations are also practicable)	4 Multi	Yes, with second frequency	22–23vdc Transmitters 12 amps Receivers 7 amps
Toran	2 or 4 1600–3800	0.4 to 0.9	CW	50–750	Hyperbolic	Multi	Yes, with additional frequencies	150–444w at 12vdc (shore) 96–288w at 12vdc (ship)
Toran 0	1 1600–3000 (Uses atomic frequency standard for stability)	0.4 to 0.5	CW	1000	2 Range or 3 Range	Multi	No	336w at 24vdc (shore) 72w at 24vdc (ship)
Raydist	2 1600–4000	0.4 to 0.9	CW	120–400	2 Range or Hyperbolic/ Elliptical	Multi Multi	No	60w at 24vdc (shore) 168w at 24vdc (ship)

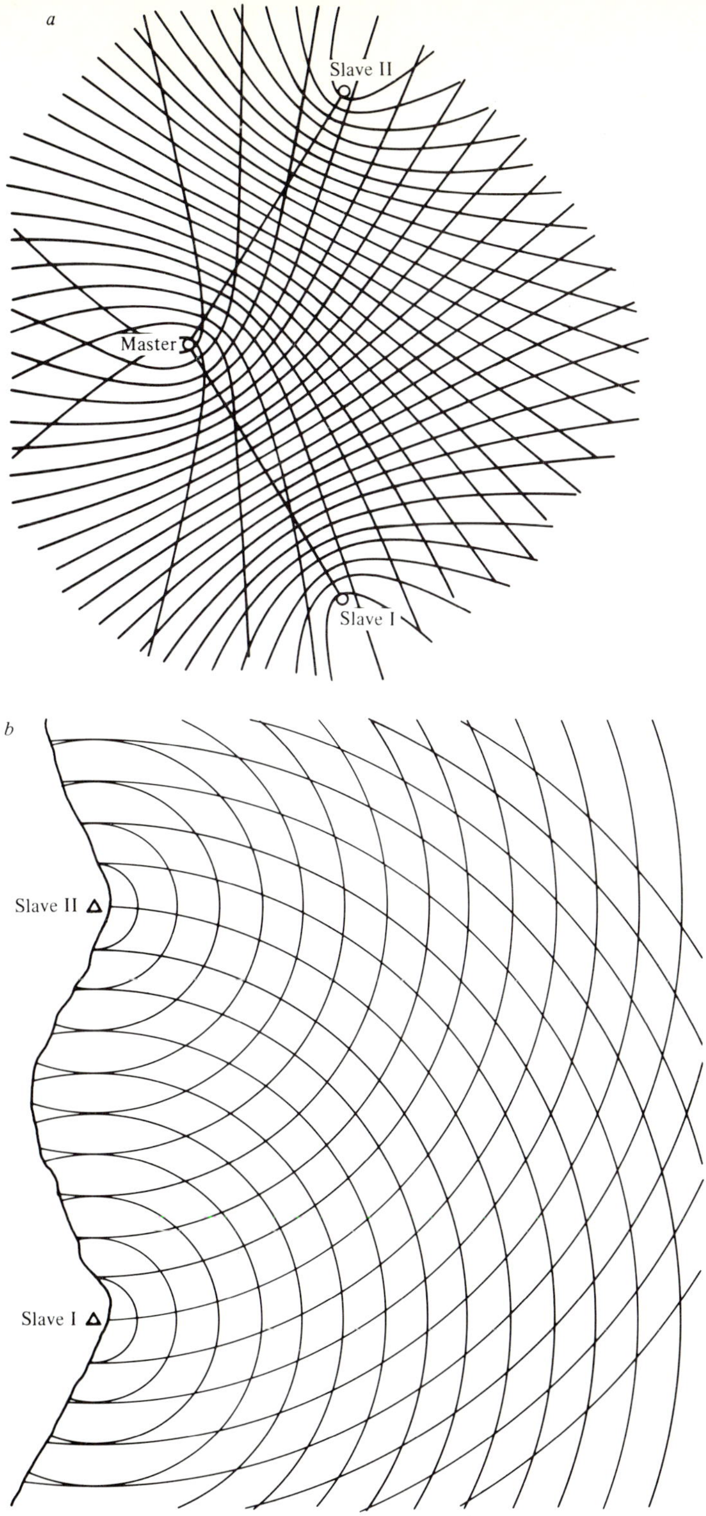

Fig. 3.4. *a*. Typical 2 mHz system – hyperbolic mode; *b*. Typical 2 mHz system – 2 range mode.

Table 3.4. *Characteristics of the main microwave positioning systems*

System	Frequency (MHz)	Resolution (m)	Nominal accuracy (m)	Type of transmission	Nominal maximum range (line of sight) km	Power requirements	Maximum number of remotes	Maximum number of users
Tellurometer ME	2800–3200	0.1	± 1	CW	50	12vdc 48w (ship) 32w (shore)	1 per ship station	1
Trisponder 202	9200–9500	1.0 (or 1.0 foot)	± 3	Timed coded pulse	75	24dc 50w (ship) 30w (shore)	4	4
Autotape	2900–3100	0.1	± 0.5 $\pm 1:100{,}000$	CW	100	12vdc 100w (ship) 75w (shore)	—	—
Mini Ranger	5450–5600	1.0	± 3	Timed coded pulse	40	24vdc 150w (ship) 13w (shore)	4	1

The Decca Trisponder 202 and Motorola MiniRanger measure the two-way travel times of coded pulses between the shipborne Master and the shore Remote station and convert these times into distance. These systems are unambiguous but require prior calibration before being put into operation, to take account of built-in delays in the Master and Remote units. The calibrations should be checked periodically, and certainly at the completion of operations.

Track control

Almost all marine electronic position fixing systems are compatible with some form or other of track-plotting device to give a continuous pictorial display of the ship's position. Some manufacturers offer a purpose-built unit, while others offer an interface for operation via a programmable calculator or computer.

Combined system

It is of course quite practicable to use any combination of the methods and systems described to position the survey vessel. One combined method frequently used is theodolite bearing and Tellurometer MRB 201 or other microwave range measuring device. In this method it is possible to 'talk' the vessel along a predetermined bearing while measuring ranges periodically, either at the vessel or on shore depending upon the disposition of Master and Remote stations.

Depth

Measurement of depth by echo-sounder

Echo-sounders measure the two-way travel time of a pulse of acoustic energy from the vessel to the sea-bed and back, and convert this to water depth. Survey echo-sounders usually employ a belt- or radial-arm-driven stylus passing across a strip of specially prepared paper to provide a continuous profile of the sea-bed immediately beneath the survey vessel. They employ a higher stylus speed, and hence faster pulse repetition rate, than conventional navigational echo-sounders in order to provide a maximum of cover over the sea-bed and a comparatively large vertical scale.

Typically, transmission frequencies lie in the band 30 to 210 kHz. Lower frequencies are used in deep water because of the attenuation of the high frequencies.

For accurate depth measurement it is necessary to be able to adjust the stylus speed according to the prevailing velocity of sound in water, and to be able to set the transmission zero so that the zero of measurement is the

Table 3.5. *Characteristics of some typical survey echo-sounders*

Instrument	Manufacturer	Frequency (kHz)	Beam width to half-power points (degrees)	Approximate maximum range (m)	Range scales (m)	Paper speeds	Tachometer or calibration mark?	Resolution
MS48	Kelvin Hughes Division of Smiths Industries Ltd (UK)	30	17° fore and aft 25° athwartships	1000+	4 basic ranges: 20, 40, 80, 200 metres, each with 6 increments (or these ranges × 2)	Variable according to range, between limits of 7 and 280 mm/min	Tachometer	10 mm on shallowest range
Fathometer Type DE719B	Raytheon (USA)	200	8°	120	0–16.5 15–31.5 30–46.5 45–61.5 (or these ranges × 2)	1, 2, 3, 4 inches per minute	Calibration mark	0.5% of depth
DESO 10	Krupp-Atlas Elektronik (W. Germany)	Any two from 30 33 80 100 210	Fore and Aft / Athwart 20° / 30 [a] 19° / 29° [a] 15 / 20 15° / 15° 8° / 8°	1400	*GPA* 0–280m Basic ranges 20/40/80m Phased in 20/40/80m stages *GPB* 0–1400m Basic ranges 100/200/400m Phased in 100/200/400m stages	1.25[b] 2.5 5.0 10.0 cm/min	Tachometer	29m range ± 5cm 40m range ± 10cm 80m ± 20cm

a Narrower beam available with different transducers.
[b] Or true-to-scale with special navigation equipment.

sea surface. The characteristics of some typical echo-sounders are summarised in Table 3.5.

Calibration

A survey echo-sounder should be calibrated at least at the beginning and end of survey operations and whenever a change in the velocity of sound in water is suspected. In estuary depths it is usual to employ the standard 'bar check' to calibrate a survey echo-sounder.

For bottom-mounted transducers the bar is in effect, just that; a bar of angle iron, or scaffold tube sealed at its ends to retain air inside, suspended horizontally beneath the transducer on two graduated lowering lines (Fig. 3.5).

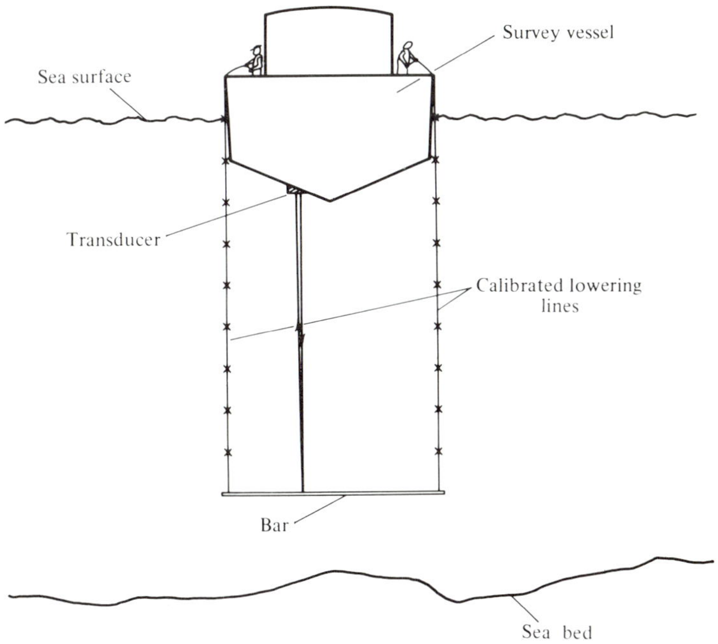

Fig. 3.5. Bar check for bottom-mounted transducers.

In the case of a temporary installation, with the echo-sounder transducer over one side of the survey vessel, a perforated metal disc would be used instead of the bar.

The 'bar' is lowered successively in, say 2-m steps, beneath the transducer, down to a maximum depth of about 20 m. The corresponding depths are recorded and read off from the paper record and plotted as errors against depth. The errors are then adjusted out by using a combination of stylus speed and zero settings.

Raising the bar in successive steps from the maximum depth used will confirm that the correct adjustments have been made. Alternatively the marks of the bar on the record can be used to construct a scale.

Survey echo-sounders incorporate a stylus motor tachometer as a calibration mark on the paper and this, combined with the zero mark on the paper, permits a constant check to be kept on stylus speed and zero settings.

Datum levels

As the zero of depth measurement in a properly adjusted survey echo-sounder is the surface of the sea, it follows that the depths recorded will vary according to the height of the tide at the time. For this reason, all depths must be reduced to a constant datum level. Sounding datum normally approximates to a low-water level, but sometimes coincides with the local land levelling datum in projects related to civil engineering works.

Tide-level readings must therefore be applied to all recorded depths before plotting (Fig. 3.6). It may be necessary to make time and range corrections

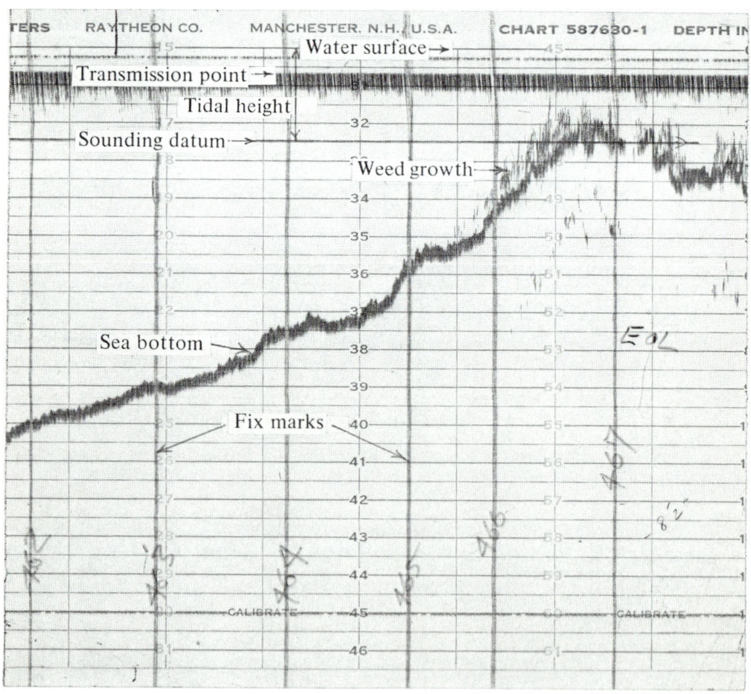

Fig. 3.6. Example of echo-sounder record.

to tide readings if they are taken at a gauge some distance from the survey area. In estuaries used by commercial shipping it may be assumed that suitable datums, tide gauges and correction factors have already been established and should be adhered to unless there is some good reason for departing from them.

Interpretation of echo-sounding records

There are a number of factors that must be borne in mind when reading the depth off the echo-sounder. Most of these are associated with fake echoes and of themselves can provide useful ways of studying other estuarine phenomena.

Side echoes

Because the beam of the echo-sounder spreads out as a cone, the most direct sound path to the sea-bed may not be vertically under the transducer. Navigationally this presents no hazards as the minimum depth will result and this is what is required. However if one is interested in the roughness or slope of the sea-bed then errors can be introduced. The best solution to this problem is to get the echo-sounder transducer as close to the bottom as possible.

Boat wakes

Air bubbles in the water will resonate to the sound pulse if their dimensions are critical. The acoustic energy is thus reflected and dissipated, and the sea-bed return is lost. The bubbles occur in the wakes of boats and as it only takes a few bubbles per cubic metre to blanket out the signal, the effect can persist some time after the boat has passed.

Fish and seaweed

Fish with gas-filled swim bladders also cause reflection of the acoustic pulse and appear as mid-water echoes. This has been the subject of considerable research and is utilised in special 'fish-finding' echo-sounders. Seaweed causes a very similar effect (Fig. 3.6), especially those species with gas-filled bladders.

Fluid mud

Ghost echoes are often observed above the sea-bed in estuaries. These are produced by layers of fluid mud; high concentrations of suspended sediment (see p. 173). Often the layering can be observed at one frequency but not at others. No general rules are yet available to explain fully and use quantitatively the phenomenon, but echo-sounding can be a useful descriptive tool in examining suspensions (see, for example, Prioni *et al.*, 1975).

Salt-wedge profiling

Reflections can also be obtained from the top of the salt wedge (Fukushima *et al.*, 1964; Fukushima, Kashiwamura & Yakuwa, 1966). The frequency used is rather higher than conventional echo-sounders and it is not clear whether it is the density step or possible organic debris floating at that level which is responsible for the reflections. However the technique has considerable potential in estuarine research.

REFERENCES

Fukushima, H., Kashiwamura, M., Yakuwa, I. & Takahashi, S. (1964). A study on salt water wedge at river mouth by ultrasonic method. *Coastal Eng. Japan*, **7**, 101–7.

Fukushima, H., Kashiwamura, M. & Yakuwa, I. (1966). Studies on salt wedge by ultrasonic method. *Amer. Soc. Civil Eng.* Proc. 10th Conf. Coastal Eng. 1435–47.

Prioni, J. R., Rona, D. C., Lanter, C. A. Jr. & Sellars, R. L. (1975). Acoustic observations of suspended particulate matter in the ocean. *Nature, London*, **254**, 413–15.

GENERAL READING

Admiralty Manual of Hydrographic Surveying, NP 134, 2nd Edition. Volume 1, 1965. Volume 2, 1969–73. Ministry of Defence, Hydrographic Department, Taunton.

Admiralty Tidal Handbook, NP 122. Volume 1, Admiralty semi-graphic method of harmonic tidal analysis, 1959; Volume 2, Tidal datums for hydrographic surveys (and other related subjects) 1975. NP. 122(2); Volume 3, Harmonic tidal analysis for short period observations, 1964. NP 122(3).

Ingham, A. E. (ed.) (1975). *Sea Surveying*. John Wiley & Sons, Chichester.

Margrett, A. *Hydrographic surveying for development and conservancy*. Kelvin Hughes & Sons, London.

4

Side-scan sonar and reflection seismic profiling

B. D'OLIER

Side-scan sonar

In the latter half of the nineteenth century and in the early years of the twentieth a great deal of fundamental research into the elastic properties of various media was carried out. In particular, experiments demonstrated the ability of water to conduct mechanical vibrations, and showed that this energy will be reflected or refracted at any boundary separating two media of contrasting properties, e.g., seawater and the sea-bed–sediment interface. R. A. Fessenden in 1912 was the first person to apply these principles when he built an apparatus to measure the time taken for sound waves to travel to the sea-bed and back, and later developed it for 'secret' communication underwater from ship to ship using Morse Code.

From experience gathered during the First World War arose the need for increased accuracy of water-depth measurement and, closely coupled with this aim, a means of detecting objects underwater, be they wrecks, shoals, rock bars, fish or submarines. During the 1920s the first practical echo-sounder using high frequency sound waves was produced. It was soon realised that by turning the transducer into the inclined or even the horizontal mode an effective means of detecting underwater objects had been created. Development continued, resulting in equipment used so effectively to detect submarines during the Second World War. This equipment was called ASDIC, an acronym for Anti (Allied) Submarine Detection Investigation Committee. Many ASDIC operators soon realised that different sections of sea-bed over which they were passing produced return signals of differing intensity. This observation was followed by work expressly aimed at mapping the sediments and features of the sea-floor (Chesterman, Clynick & Stride, 1958). In 1956 Knott & Hersey reported on work involving a pair of towed transducers set to examine the sea-floor on both sides of a ship at the same time. The first purpose-built geological side-scan sonar (Sound Navigation and Ranging) was manufactured to a British National Institute of Oceanography design in 1960 and a version was made available commercially by Kelvin Hughes Ltd in 1962. Since then development has been extremely rapid and the side-scan sonar has become a standard and essential tool for the geological study of any lacustrine, fluvial, estuarine or marine environment.

Principle of operation

The principle of operation of a side-scan sonar is essentially similar to an echo-sounder. A pulse of electrical power is applied via a switching unit to the transducer causing it to vibrate, thereby creating pressure waves which are projected into the water. On reflection from the bed of the estuary, the pressure waves are picked up by the receiving transducer or hydrophone, converted into electrical energy and usually recorded on a continuously running paper recorder (Fig. 4.1).

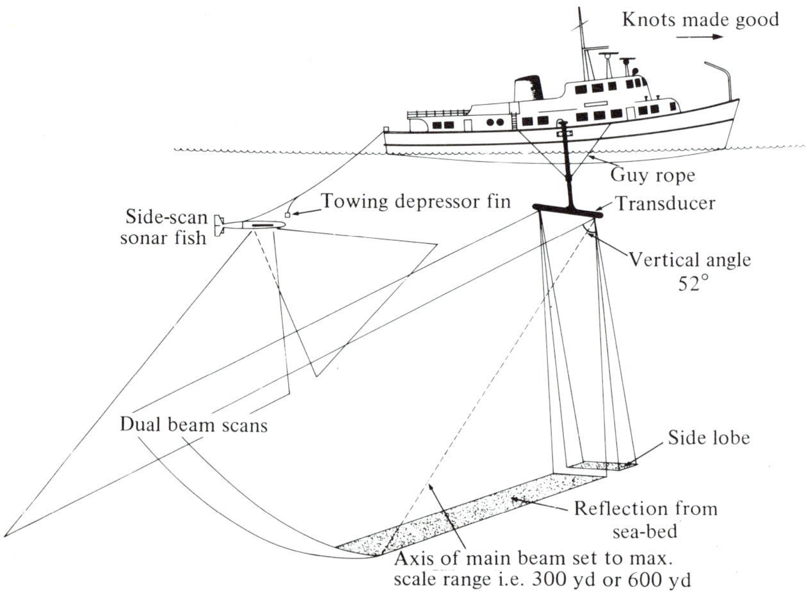

Fig. 4.1. Diagram showing principle of operation of hull-mounted and towed dual-channel side-scan sonars.

However, because the acoustic waves are produced with similar frequency and wave-form to the electrical pulse, it is important that the pulse is of uniform power and duration and matched to the resonant frequency of the transducer. In simple, relatively unsophisticated sonars, electrical energy is passed by contact points to the transducer. This however results in acoustic waves of variable power and frequency being emitted. In the more advanced sonar designs the electrical pulse is obtained from continuous wave generators and controlled by tuning fork or crystal. This ensures that a precise frequency is being passed to the transducer, with the switch unit passing pulses of the required duration and interval.

The transducers used are of three basic types:

(a) Electrostrictive. Materials such as barium titanate, after being electrically polarised, alter in their physical dimensions in proportion to the magnitude of an applied charge.

(b) Magnetostrictive. Materials such as permanently magnetised nickel can be made to expand and contract in the presence of a fluctuating magnetic field.

(c) Piezoelectric. When alternating current is applied to quartz and other crystals, a change in the dimensions of the crystal will occur with a resonant frequency of vibration inversely proportional to its thickness.

Transducers used for side-scan sonar operation can either be fixed to the hull of a ship where they generally scan to one side of the vessel's track only, for example, the Kelvin Hughes MS 47, or they can be towed in a fish, the latter usually being a dual-channel system allowing scanning to occur on both sides of the ship's tracks, e.g., the EG and G 1B (Fig. 4.2).

The transducers are generally made to vibrate within a frequency range of 5 kHz to 300 kHz and most of those commonly used are of the piezoelectric

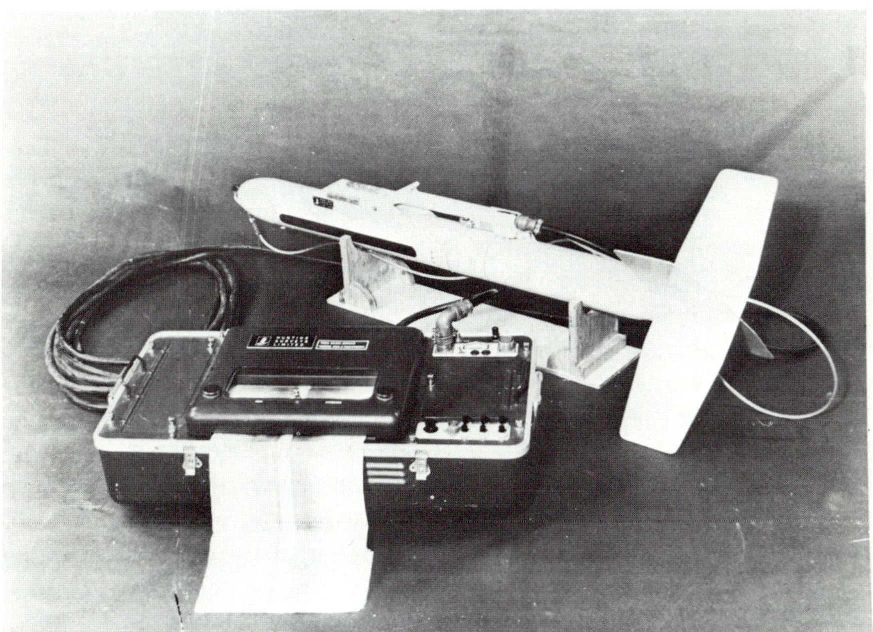

Fig. 4.2. EG & G dual-channel sonar with towed transducer. (Photograph: courtesy Huntec Ltd.)

type. The early models tended to use the magnetostrictive principle of operation and operated most efficiently at frequencies below 50 kHz.

However as was realised early in the development of acoustic apparatus for underwater use, the choice of frequency has important implications concerning the efficient use of the apparatus. Acoustic energy is lost through the spherical spreading of power away from the source and by absorption due to the conversion of acoustic power to heat. Losses due to absorption increase linearly with range and are frequency dependent. The higher the frequency of the acoustic waves the more rapidly they are attenuated in water. For example, whilst a high frequency of 300 kHz might be almost completely attenuated after travelling 100 m, an acoustic wave of 50 kHz would have only suffered some 30% loss of power, whilst an acoustic wave of a frequency of 5 kHz would have only lost some 8%. It follows therefore that high frequency transducers are generally used for shallow water and medium depth echo-sounding and for short range side-scan sonar, whilst low frequency transducers are used for sediment and rock penetration studies, for deep water echo-sounding and long range sonar.

Resolution

The resolution or clarity of the returned and recorded signal is dependent upon a variety of factors, all of them important in determining the capability of equipment to measure accurately distance and size of underwater objects, to record their shape and to distinguish between them when they are in close proximity.

Directivity and beamwidth

Transducers are designed to send out beams of sound of various shapes, although, usually, a level of directivity is required so that accurate measurement of sea-bed features may be achieved. To have an acoustic beam narrow in the horizontal dimension is important otherwise echoes might be received at one time from features lying ahead, astern or on either side of the vessel, whilst a beam wide in the vertical dimension is important so that scanning might be achieved from directly under the vessel to an adequate range on either (single channel) or both sides (dual channel). Directivity is related to wave length and therefore to the frequency of the transmission and also to the size of the transducer. The degree of directivity is indicated by the transducer's beamwidth. This is defined as the angle between the points at which the sound energy intensity has fallen to half of that along the beam centre axis. For a rectangular transducer the beamwidth is given approximately by the expression

$$B^{o} = \frac{50\lambda}{L} \tag{4.1}$$

where B^o is the beamwidth, λ is the wave length of the transmitted frequency and L is the dimension of the acoustic wave transmitting surface at right angles to the plane of the beam being considered. For example, assuming that the velocity of sound in water is 1520 m s^{-1}, and that the transducer emits an acoustic wave of frequency 48 kHz, the wave length of that transmission will be (1520 ÷ 48 000) m or 0.0316 m. The beamwidth of a rectangular transducer of dimensions 1 m × 0.03 m will be

$$\text{horizontal dimensions} \quad \frac{50 \times 0.0316}{1} = 1.58^o,$$

$$\text{vertical dimension} \quad \frac{50 \times 0.0316}{0.03} = 52.67^o.$$

Directivity can be improved by placing reflectors, usually materials of high porosity and frequently of high air-bubble content, around the transducer, thereby focusing the acoustic vibrations into accurately pre-determined beam paths.

Pulse length

In order to distinguish two closely spaced objects the return signal from the first should not mask that from the second. To achieve this the targets should be greater than half a pulse length apart, where pulse length is defined as pulse duration × velocity of propagation. A switched or gated pulse of 48 kHz emanating from a continuous electrical wave generator and lasting one millisecond will consist of 48 cycles. With a sound velocity in seawater of 1520 m s^{-1} and a resulting wave length of 0.0316 m, the pulse length will be $48 \times 0.0316 = 1.52$ m. The objects must therefore be more than 0.76 m apart in order to be resolved separately.

Recording sensitivity

However sophisticated transmission systems are, the effectiveness of operation must depend finally on the quality of signal recording. To accommodate the wide and variable range in intensity of the returning signal, a recording system with a high dynamic range is required. Records can be made on 'dry' or graphite backed paper. Here, the return signal is earthed to a platen lying behind the paper, so that the facing paper is burned off and the graphite is exposed. However, the process of burning is not especially sensitive to variations in the return signal strength and consequently this type of recording paper is only used for general surveys. Although increasingly sensitive 'dry' papers are being developed resolution is generally only possible for features larger than about 0.75 m. Wet or chemically treated paper, which reacts under the stimulus of an electrical discharge to turn black/grey or sepia coloured, is far more sensitive to variable return signal strengths and can resolve features as small as 0.30 m. However, the paper often distorts on drying and will fade if exposed to sunlight.

To achieve even better results the returning signals can be taped and the recording played back on to a moving paper recorder using a selection of frequencies, picking the frequency range that provides the best and clearest record of the estuary bed.

Sonograph interpretation

The acoustic signal once transmitted is dissipated or attenuated as it travels through the water, by spreading and by absorption. The longer the signal path the greater the loss. This is of great importance since the return signal strength and consequently the recording density or tonal variation on the recording paper is used to provide evidence of lateral change in sediment type and morphological feature. To compensate for this the recording apparatus usually incorporates a 'time varied gain', the amplification of the received echoes being increased with the greater distance of their point of origin from the ship.

Once the signal strikes the estuary floor further losses occur. The process is a complex one involving the characteristics of the sediment, the roughness and slope of the sea-bed and the angle of incidence of the acoustic energy. The greatest proportion of the energy is reflected, a small amount penetrates into the sea-bed to be reflected and refracted by other discontinuities, and some is scattered. Direct reflection is only likely to occur close to the boat, elsewhere the back scattering energy will be detected. The reflectivity (R) of the bottom for normal incidence is given by

$$R = \frac{Z_2 - Z_1}{Z_2 + Z_1} = \frac{A_R}{A_I}, \qquad 4.2)$$

where A_R = Amplitude of the reflected signal
A_I = Amplitude of the incident signal
Z_1 = Acoustic impedance of water = $d_1 v_1$
Z_2 = Acoustic impedance of bottom material = $d_2 v_2$
d = density
v = velocity of propagation.

For a back scattered signal the amplitude depends on the size of the back scattering elements and the angle of incidence. The relationship is not fully explained but in general the reflectivity is directly dependent on sediment type, the coarser the sediment the greater the reflectivity and consequently the tonal intensity of the sonograph. It is of great significance, however, that the acoustic impedance bears a broadly inverse relationship to porosity; high porosity sediments have a low impedance and therefore a low reflectivity whereas low porosity material has a higher impedance and reflectivity. High porosities tend to occur within fine-grained sediment and lower porosities in the coarser-grained materials. Sometimes there is a great deal of similarity between porosities in different sized material as porosity does not entirely depend on grain size. Factors such as grain shape, the degree of sorting, the length of time since deposition and therefore the degree of settling and compaction, must all enter into the reckoning. Clays might have porosities ranging from 65–82%, silts 45–88%, fine and very fine sands 40–50% and medium sands 37–42% (Taylor Smith & Li, 1966). Materials coarser than medium sand, can if 'clean' have as high a porosity as some silts but, because of their capacity for trapping finer material within their pore spaces, the range of grain sizes within a deposit of this sort will affect sonograph tone intensity. If the particle size exceeds the wave length of acoustic radiation very high reflectivities result, although a sample might show that the interstices are packed with a clay or silt matrix and only the surface washed clean by the tide is of well sorted material. Sands, particularly those of fine–medium grain size which have been deposited quickly, have a loose unstable packing arrangement. By slow deformation, usually brought about by slope creep or by the shearing action of tidal currents and by pressure fluctuations resulting from wave action, a more compact, stable, grain arrangement results. Therefore porosities and consequent reflectivities might well vary where very recently deposited sediments are being studied. Reflectivity is also affected by the angle of incidence of the sound pulse to the estuary floor or bottom feature. Topographic slopes facing the outgoing acoustic pulse will give a strong reflection whilst slopes facing away from the pulse, if not so steep as to be in shadow, will give a poor reflection. Acoustic pulses with low grazing angles to the estuary floor are largely reflected away from the ship.

The acoustic pulse might be reflected not only from the bed of the estuary but also from an interface below it. This is most likely to happen where the angle of incidence is high, when the upper layer has only a small velocity and density contrast with the waters of the estuary and when its upper surface is much smoother than its lower surface. Also, if the operating frequency contains an appreciable element of very low frequency energy, penetration might occur. In estuaries, for instance, mobile 'pools' of fluid mud containing very high percentages of water have been observed, sometimes as thick as 2 m. These 'pools' are readily penetrated by the acoustic signals of most side-scan sonars. Very high frequency equipment is necessary to map confidently areas where this condition is prevalent, for example in deep holes in channels and the quieter harbour areas. This will improve the back scattered energy received from the upper interface and decrease that received from the lower interface because of attenuation. Careful sediment sampling, observation using underwater television or inspection by divers may be necessary where these conditions are suspected.

Ambient noise

Sometimes the signal recorded on the recording paper has a 'fuzzy' or ill-defined quality. This may be due to several factors including interference from other ship-borne equipment or from faults within the equipment itself. However, one of the most common causes is the swamping of the return signal by sound of a wide frequency range, emanating from nearby, but foreign sources – the 'ambient sound'. This might consist of the sound of a passing ship's engines, the breaking of waves at the water surface or on a nearby sandbank or shore. Other causes are the grinding together of boulders and gravel in shallow water under the influence of a heavy swell, the sound of rain on the water surface, biological noise caused by shrimps and other small creatures, or even the too-friendly attentions of a school of squeaking dolphins. Ambient sound can also be picked up from the crests of sandwaves due to the movement of sand over the ridge crests at certain states of the tide.

Sound scattering

The clarity of sonographs and the effective range of a sonar can be seriously affected by the scattering of sound from objects or surfaces dispersed within the water, which have dimensions smaller than the wave length of the sound pulse. This scattering and reverberation is generally caused by air bubbles. The effect of a ship's wake and the action of the propellers can be striking in obliterating part of the sonar record, particularly if the vessel is travelling fast and light. The ship leaves elongated ribbons of bubbles that sometimes persist until it has dipped over the horizon. It only needs one bubble 1 mm in diameter per cubic metre of water to blanket the signal. Similar, but total, loss of signal can occur by quenching on the face of the transducer either by towing too fast or, in the case of hull-mounted transducers, by the vessel rolling. An occasional occurrence can be the echoes received from the swim bladders of fish either swimming in shoals or lying on the estuary floor. Generally these do not seriously affect range but can be hard to distinguish from rock outcrops and other features.

Shadow zones

Other problems of interpretation can be caused by variations in the velocity of sound in turn caused by temperature, salinity and pressure changes with depth of the water. The velocity of sound increases at 4.5 m s^{-1} $°C^{-1}$, at 1.3 m s^{-1} $‰^{-1}$ and at 1.7 m s^{-1} 100 m^{-1} increase in depth. The equipment can usually be adjusted for changes in velocity of sound where these can be predicted but difficulties arise in areas of mixing, particularly where waters of widely differing temperatures are meeting. Due to refraction, a sound beam will be split where a cool layer underlies a warm and consequently higher velocity layer. Part of the ray is reflected towards the bed of the estuary and part is reflected to the surface leaving a shadow zone at intermediate ranges. Complex patterns, often appearing as ripple-like features on the sonograph and produced by temperature refraction can be misleading. Salinity variations, particularly at times of high river flow and at all times within the mixing zone of estuaries, can cause similar but usually less marked effects.

Lloyd's Mirror effect

Another false trace frequently met with in shallow water conditions is the Lloyd's Mirror effect. In exceptionally calm weather when surface water waves are at a minimum, the water surface can act as a mirror. Reflectivity of a sound pulse at the air–sea interface is very close to unity resulting in almost perfect reflection. At the same time a phase reversal occurs at the surface. Consequently the returned signal reflected from the water surface on to the transducer interferes with the direct return, resulting in the target signal

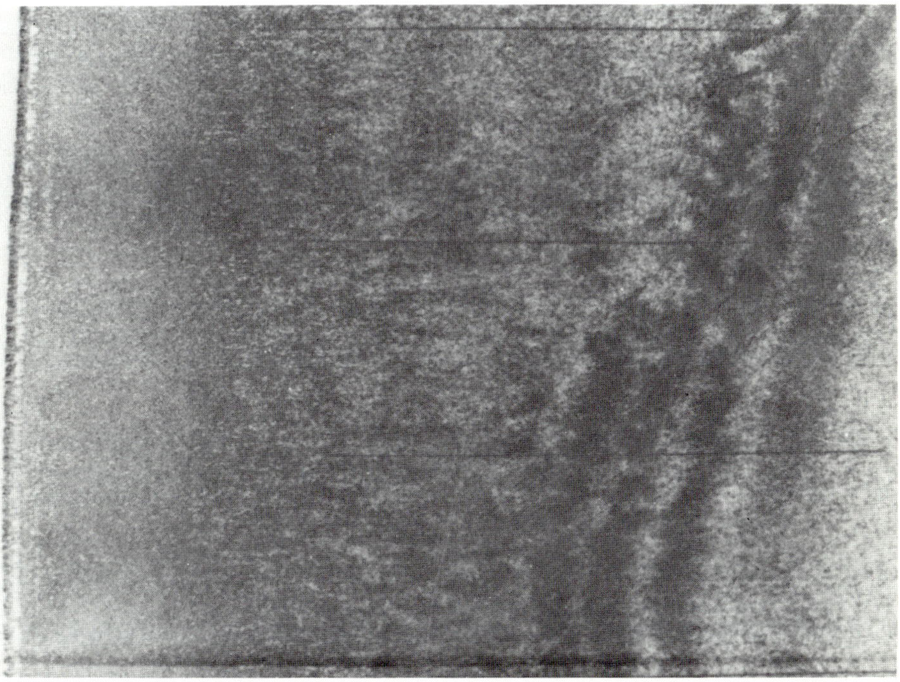

Fig. 4.3. Single-channel sonograph showing Lloyd's Mirror effect.

alternately doubling and cancelling. This results in dark and light interference fringes – the Lloyd's Mirror bands, running approximately parallel to the ship's track, across the sonograph (Fig. 4.3). The effect is common with hull-mounted transducers and can be reduced by tilting the transducer from a horizontal plane, so that the signal reflected from the water surface is received at different times along the transducer's length. With towed fish transducers it is less common as generally only the weaker side lobes of the sound beam approach the surface; deeper towing of the transducer usually dispenses with the mirror fringes. Lloyd's Mirror can be a useful effect, however, as variations in water depth and topography can be measured, since the width and separation of the Lloyd's Mirror bands is related to the water depth.

Survey accuracy

The accuracy and completeness of a side-scan survey depends on a wide variety of factors including vessel speed, the pulse repetition rate, the effects of pitch, roll and yaw, as well as the correct handling of the equipment and accurate navigation. An isometric or true plan record, where distance across the record is equal to distance along the sonograph, can be obtained as long as the vessel's speed over the ground is closely matched with the paper speed of the recorder. This is difficult and unlikely to be maintained for long. To achieve truly isometric sonograph records, then, either specialised photographic techniques need to be employed, or the data must be recorded on magnetic tape, and replayed through computerised display equipment which is programmed to give isometric presentation (Kelland, 1972). At other times frequent 'fix-marking' of the records helps to provide a fairly accurate positioning of features on the estuary floor. When plotting the position of features relative to the ship's accurately fixed course, difference between the ship's heading and plotted course has to be taken into account. Also, slant range distortion may need to be considered. A feature has to be horizontally abeam of the transducer in order to be accurately positioned according to the range scale shown on the sonograph, but as the transducer is always some distance above the bed, a correction has to be applied. With towed fish systems, the transducer is usually operated at a distance above the floor equal to 10–20% of the range scale in use. Owing to the obliquity of the sonar beam equal true range intervals will not follow the linear scale on the sonograph. The short-range intervals are compressed whilst the far-range intervals are slightly extended.

The true range

$$R_T = (R_S^2 - H_T^2)^{\frac{1}{2}}, \tag{4.3}$$

where R_S = the slant range

H_T = the height of the transducer above the bed.

Flemming (1976) gives a series of nomograms to aid in this important but tedious process.

At the same time the height (H_O) of an object can be calculated from the width of its shadow zone (W_S) and the position of the object relative to the transducer,

$$H_O = \frac{W_S \times H_T}{W_S + R_S}. \tag{4.4}$$

The slowest possible ship's speed is best to achieve a clear and fully detailed sonograph, as this increases the time that any underwater objects are within the sonar beam, thus giving a more detailed picture of their shape and position. Travelling too fast might result in some objects being completely missed, particularly if the pulse repetition rate is low.

Other problems likely to be encountered which make any accurate survey difficult are the effects of pitch and yaw. The effects of these are most marked when using hull-mounted transducers; any movement of the ship from the perfectly stable mode will mean the sonograph will not be recording the estuary floor features directly abeam of the ship. When the ship is rolling heavily, quenching might occur when air bubbles in the surface water absorb and reflect the sonar beam cutting down range and effectiveness appreciably. All of these effects are less in towed fish systems, although in heavy seas highly accurate surveying is difficult as the fish will pitch, roll and yaw due to tow cable vibration and the orbital motion of passing waves. An important factor to be noted in the accurate plotting of sonar survey lines, is the effect of current drift on the ship's course. A note of both the ship's 'head' and frequent and accurate position fixing must be carried out so that the 'course made good' can be plotted and the sonar results properly related to it.

Application

For the initial exploration of the sediments within an estuary the side-scan sonar has proved to be extremely useful. However, it is generally necessary to carry out a bed sampling programme as well, so that a check can be kept on the interpretation given to the resulting recorded data. Many of the sediment types likely to be met with in this environment have distinct recorded signatures, though sometimes different bed conditions will give similar records. For example hard homogeneous silty clays, such as the London clay, have a similar recorded appearance to the overlying lag gravels. Generally in this clay a hard band of sandstone will eventually give away its identity, but the boundary between the Tertiary rock outcrop and the Holocene lag gravels cannot usually be drawn relying on the sonar alone, both appearing as a dark homogeneous tone on the recording paper. Notwithstanding these difficulties, the sonar is of vital importance in a reconnaissance survey of estuary floor sediments and from this initial survey a sampling programme and further, more detailed, sonar work can be planned.

Where consolidated rocks have distinct bedding planes, joint sets, faults or folds and where no recent sediments thickly blanket them, the sonar will accurately record these features and their trends. It is therefore often worthwhile carrying out a sonar survey in areas of hard rock outcrop either before or during any sub-bottom profiling programme that is attempted, as it can usefully aid and supplement the latter. Besides mapping the broad distribution of the various sediment types in the estuary the sonar is most useful in delimiting areas where sediment is suspected of being very mobile. Ripples, megaripples, sandwaves, sand patches and sand ribbons are all thought to be bed-forms indicative of bedload sediment transport paths (Fig. 4.4) and attempts can be made to quantify this by repeated sonar surveys (Langhorne, 1973). Using high precision electronic position-fixing equipment, 'on the bed' transponders or even fixed stakes or concrete blocks that are big enough to be resolved and recorded on the sonar recording paper, the relative movement of the bed forms under the influence of tides, surges and waves can be carried out. Of great value in the interpretation of many types of sea and estuary floor features is the book *Sonographs of the Sea Floor. A picture atlas* (Belderson *et al.*, 1972).

The commercial applications of side-scan sonar are of increasing importance. Perhaps its foremost value in estuaries up to present has been to locate sites of valuable deposits of sand and particularly gravel for eventual use by the construction industry. Coupled with a sub-bottom profiling survey the link between the deposits likely to be of commercial value and events and processes operative during the recent geological past can be shown. For instance many of the gravel deposits can be linked with the now drowned courses of rivers, with glacial outwash spreads and with ephemeral beaches now also drowned by the recent Flandrian sea-level rise. In addition, port authorities are finding the equipment of increasing value in locating wrecks, anchors and other valuable or dangerous sea-bed obstructions. Where dredging is attempted either in or close to dock areas or in the approach channels to dock complexes, side-scan sonar can be used to monitor changes in sea-bed morphology and sediment type and to give warning of areas where maintenance dredging is likely to be necessary to keep the port approaches open. Recently a trial dredge for a projected shipping channel was carried out within a large sandwave field in the southern North Sea–Thames Estuary approaches. This has been surveyed on numerous occasions since, in order to monitor the progress of refilling that might be expected in such an area

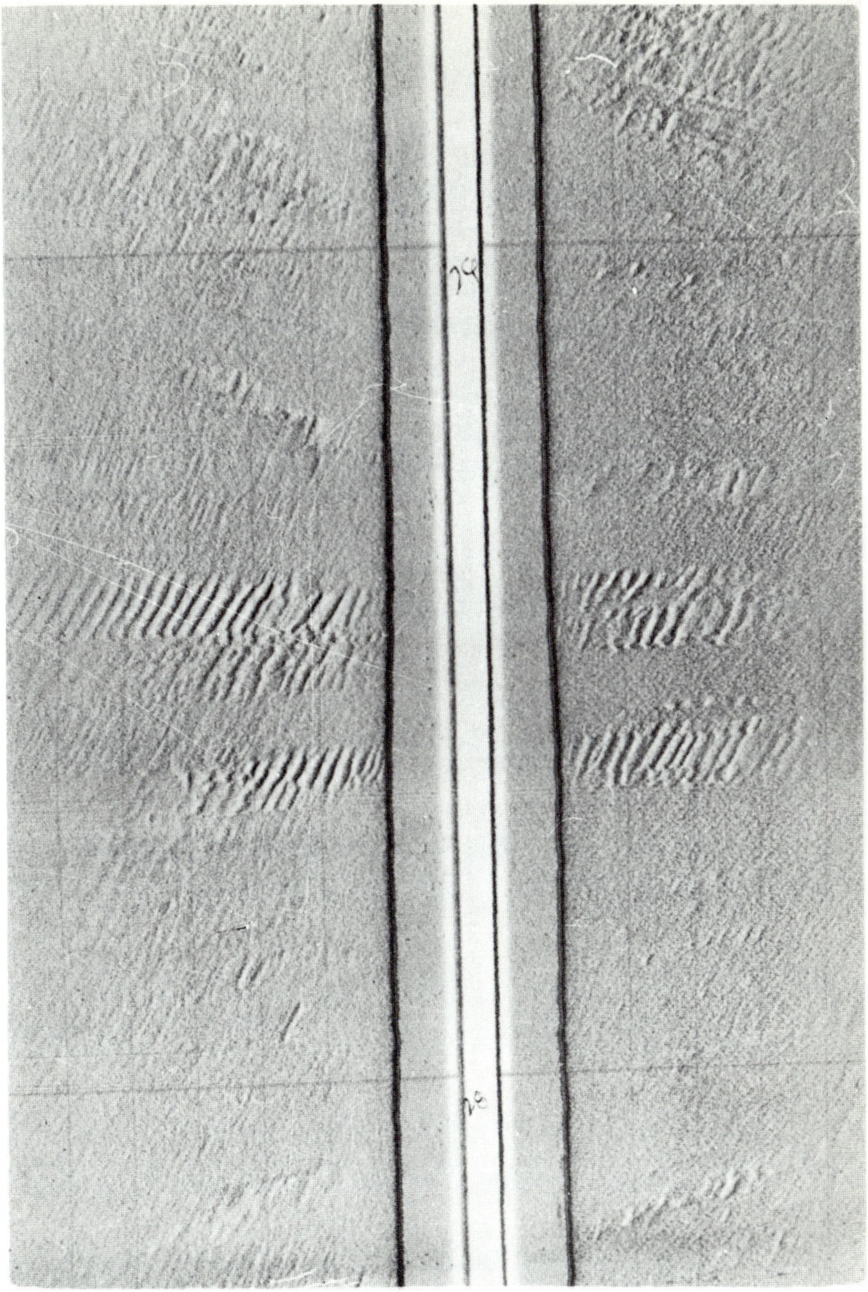

Fig. 4.4. Dual-channel sonograph showing sand ribbons.

of highly mobile sediment. In addition, useful information is likely to accrue as to the regenerative process of sandwaves once they have been disturbed or demolished. With the advent of deeper draught vessels this is likely to be a study of great navigational and commercial significance. Other uses have been in surveying pipe-line and cable routes and the sites for the location of power-station cooling-water inlet pipes. In areas of high sediment mobility the pipes and cables can become exposed and damaged and inlet pipes become covered and blocked. Recently exposed man-made obstructions might also become a major hazard for bottom-trawling fishermen.

Detailed maps of estuaries and the sea-bed of the continental shelf in many areas have been drawn up indicating morphology, sediment type, dominant and large-scale bed-forms and areas of stability and mobility. The side-scan sonar has provided many of the data for this to be carried out effectively.

Continuous seismic reflection sub-bottom profiling (C.S.P.)

Principle of operation

The basis of operation of the sub-bottom profiler is essentially the same as for the side-scan sonar or the echo-sounder. A continuous series of sound pulses is emitted from a towed source and, after reflection from the estuary floor and, additionally in this case, from discontinuities within the underlying

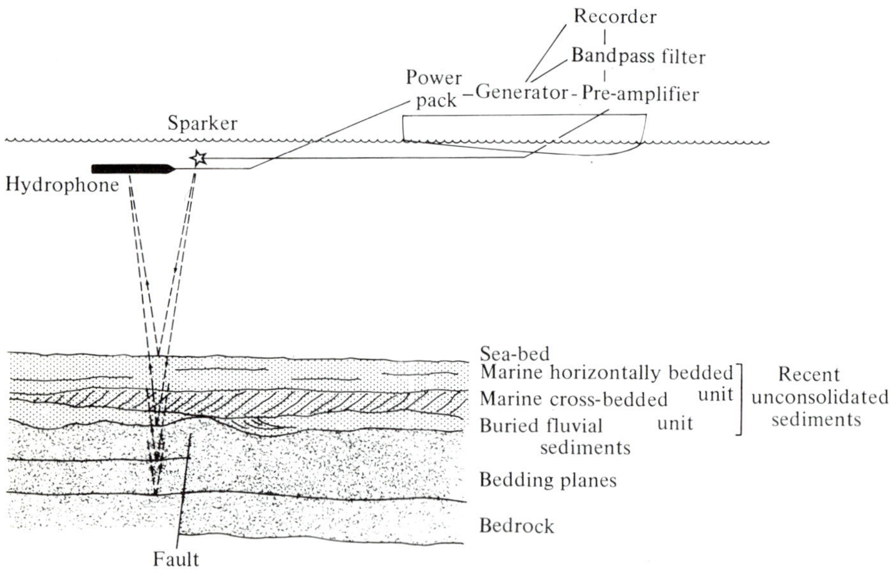

Fig. 4.5. Diagram showing principle of operation of continuous sub-bottom seismic profiling system.

sediments and rocks, the returning signals are picked up by the receiver and recorded. The time taken for the pulses to travel to these horizons and return to the detector after complete or partial reflection is recorded on a continuously running chart, forming in consequence a geological section along the survey line. The greater the depth of the reflector, the longer the two-way travel time will be (Fig. 4.5).

All seismic operations are based upon the principle that a stress (force per unit area) causes a proportional change in length or volume per unit length or volume. This proportionality is characterised by elastic moduli or constants, the two of greatest interest being the bulk or Youngs' modulus (k) and the shear modulus (n). Stress can be either compressive or tensile if it acts perpendicular to an area, or shear if it acts parallel to that area. When an impulsive stress is applied to an elastic medium, in this case water, the strain is propagated as an elastic wave. In continuous seismic profiling only the P or longitudinal/compressional waves, where the motion of the medium is in the same direction as the direction of wave propagation, are important. The S or shear waves where the particles move at right angles to the direction of wave propagation are not transmitted by water. The propagation velocity (V) of P waves of recent sediments and some common rock types are given in Table 4.1 and can be calculated by using the expression

$$V^2 = \frac{(k + \frac{4}{3}n)}{p}, \qquad (4.5)$$

where p = the density of the medium.

Table 4.1. *P wave velocities* ($m\ s^{-1}$)

	Propagation velocity
Air	330
Seawater	1450–1570
Recent estuarine sands and silts	1530–1600
Glacial till	1560–2500
London clay	1650–2100
Chalk	3000–4700
Granite	4500–6700

Velocities, particularly in sedimentary rocks, tend to increase with depth of burial and age and many attempts have been made to represent this increase empirically. Of significant importance, however, are factors such as porosity, degree of compaction, the type and extent of cementation, grain shape and mineralogy and the position and shape of grain boundary contacts. A rock with a high density and a well cemented or rigid grain structure will transmit P waves at high velocities; uncompacted, uncemented sands and silts at relatively low velocities.

Velocities can be obtained by subjecting core samples to ultrasonic impulse techniques within the laboratory. Problems arise as the action of coring, of core extraction and final sampling might well change the internal porosity and grain contact characteristics, particularly of relatively unconsolidated sediments. Where the sediments are consolidated and a drilling programme is in operation, using 'down the hole' seismometers and measuring the P wave travel time from the surface to the various sediment interface levels enables precise velocity measurements within each layer to be obtained and without the disturbance problems of cored samples. However, the interpretation of a reflection seismic profile is made so much easier by making sure the profiler survey track is coincident with a borehole or preferably a series of boreholes. The elastic velocity of each recognisable layer is then readily calculated. Nevertheless it is not always a simple matter to correlate the recorded profile reflectors with identifiable layers within the core (Fig. 4.6). Energy is reflected at discontinuities within the medium, the amount of energy depending on the reflectivity of that interface (Eqn. 4.2). Any change in the micro-structure of the sediment or rock will cause elastic wave reflection although the amount

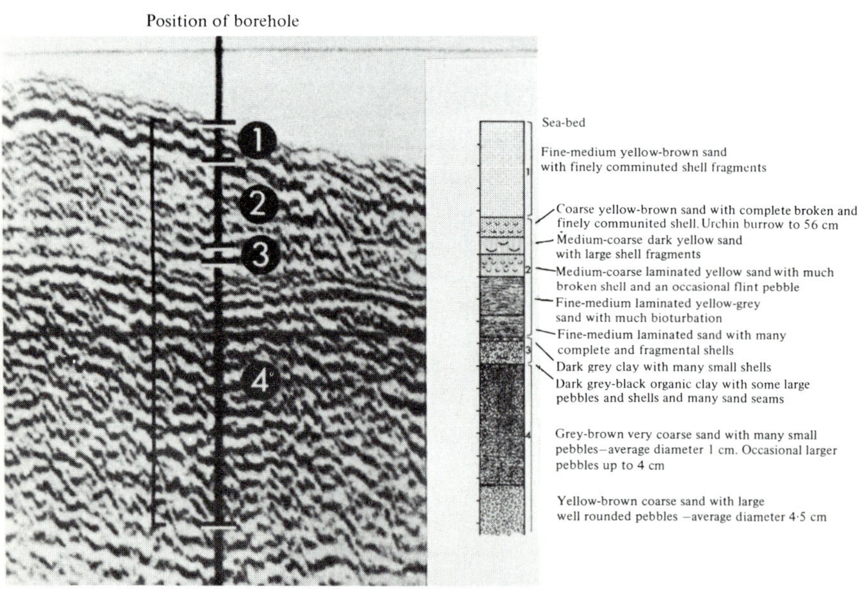

Fig. 4.6. Correlation between CSP record and borehole log.

of energy might at times be too small for efficient resolution and recording. Equally, the energy reflected at a particular layer may be great enough for resolution and recording but the change too subtle for visual identification of that layer within the core.

Before beginning a sub-bottom profiling survey it must be decided what features of the sub-surface are of prime importance within the investigation. Is it to be the broad structural features of the estuary basement rocks lying at depth below the surface, the major characteristics of the more recent overlying sedimentary sequences, or the fine detail of bedding and cross-bedding within the more superficial layers? These objectives require a slightly different approach. Sound sources used for sub-bottom profiling operate over the frequency range of 20 Hz to 14 kHz. Devices operating at higher frequencies rapidly lose most of their energy through attenuation and by reflection at the water/sediment interface and are only used for echo-sounding and side-scan sonar operation. It is found that qualities of good sediment layer resolution and deep penetration tend to be incompatible. Without sophisticated processing techniques, for example, resolution cannot be better than half the pulse length of the transmitted signal. As the pulse length increases at lower frequencies, attenuation of a sound pulse is less rapid so greater penetration results. A greater pulse length allows more power to be projected with a consequent though relatively small increase in penetration. However, too long a pulse will cause loss of resolution due to the outgoing signal interfering with the one returning. It is important, therefore, that the correct sound source is chosen with the aim of producing a short, high power pulse at a frequency range commensurate with the aims of the survey.

Seismic profiling sound sources

There is a wide range of continuous seismic profiling systems currently available including Sparkers, Boomers, Air-guns and Pingers. Table 4.2 gives the main performance characteristics of various sub-bottom profiling systems.

Sparkers. Sparkers operate by releasing electrical energy from charged condensers into the water. The stored electrical energy is discharged from either a single or a few electrodes (Sparkarray) or from multiple, very closely spaced electrodes, sometimes numbering several hundred (Fig. 4.7). When the discharge occurs water is broken down creating a gas bubble which expands rapidly to form a pressure pulse. Unfortunately, continued oscillation of the bubble, because of its adiabatic expansion and contraction, effectively increases pulse length and thereby decreases resolution.

Multi-electrode systems, because of the small discharges at each tip, have shorter pulse lengths than those with only a few electrodes, and in addition

Table 4.2. *Characteristics of sub-bottom profiling systems*

System	Frequency range (Hz)	Power output (Joules)	Resolution (m)	Penetration (m)
Sparkarray	50–2000	200–8000	5–20	50–250+
Multielectrode sparker	200–1000	200–1000	2–4	up to 150
Air-gun	20–2500	—	4–20	70+
Boomer	200–5000	500–1000	2–4	up to 200
High resolution boomer	400–14000	100–250	0.5–1.5	up to 50
Pinger	2000–14000	0.1–5.0	0.2–1.0	up to 30

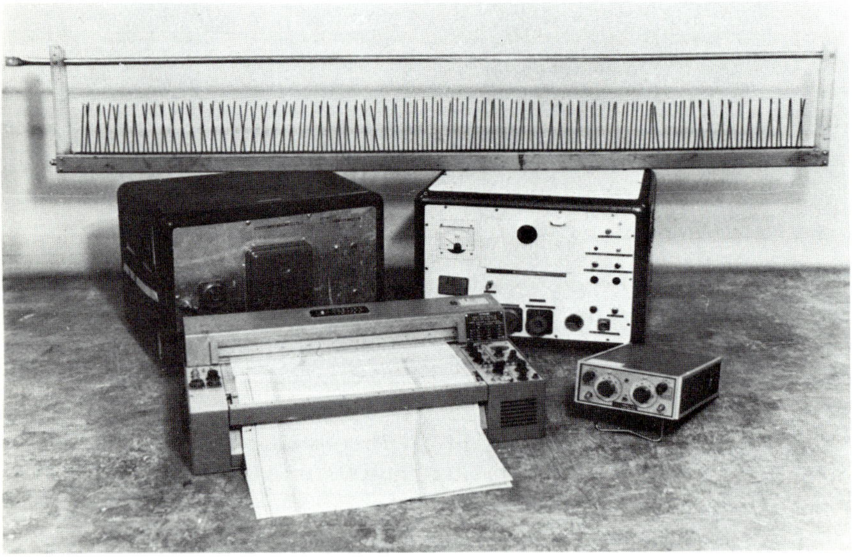

Fig. 4.7. A 150-tip multiple electrode sound source showing recorder (centre), power supply unit (left), triggered capacitor (right), and variable filter (front right). (Photograph: courtesy Huntec Ltd.)

can generate stronger pressure pulses. This modification increases both the resolving power of the system and the penetration. It is therefore a useful system for reconnoitring many estuarine sub-surface features such as sand sheet thicknesses and internal structure, buried channel macro-structure, the form of the buried bedrock surface and even such features as dip and other structural features of shallow lying basement rocks. Sparkarray systems run at low power outputs will provide a closely comparable performance, but when operated at higher power outputs with resulting greater pulse lengths they are of little value in estuaries, because of their poor resolving power. They are best used in studies of major structural features in the basement rocks and in studies of the broad detail of overlying sediments where these are of appreciable thickness.

Air-gun. The air-gun operates by the release of high pressure air directly into the water from a firing chamber. For estuarine work, a single small capacity gun operating at comparatively low pressures of 300–400 p.s.i. will produce a pressure pulse that will achieve limited penetration and resolution. The problem of an overlong bubble-pulse with consequent poor resolution has been largely overcome by modifications designed to introduce air into the bubble to cut down the continuing oscillations. Small air-guns are useful in estuaries where the aims of the survey are to ascertain the main features of the recent sediment cover and its relationship with shallow-lying older 'basement' rocks. One problem associated with omnidirectional acoustic wave sources such as air-guns and sparkers is that because the water/air interface represents a near perfect reflector a series of additional unwanted reflections arise (Fig. 4.8). Only ray path v is of importance. Owing to the time lag associated with these different ray paths the effective pulse length is increased and consequently the resolution decreased.

Boomer. A device which to a large extent overcomes this problem and obtains a reasonable compromise of penetration and resolution is the high resolution Boomer. This consists of an insulated metal plate and a rubber expanding diaphragm adjacent to a rigidly-fixed flat wound electrical coil, all of this mounted within a catamaran or towfish. When a high power electrical pulse of short duration is discharged into the coil the resultant magnetic field explosively repels the metal plate which in turns generates a single acoustic pressure pulse. In earlier designs, although greater power outputs were achieved, the plate tended to oscillate, lengthening the pressure pulse and reducing resolution. A wide spectrum of frequencies ranging between 400 Hz and 14 kHz can be generated thus allowing a compromise of penetration and resolution to be made. The frequency of operation to some extent determines the acoustic beamwidth: the higher the frequency, the smaller the beamwidth, although most of the energy is directed towards the bed with little reaching the water surface. Effective pulse length is thereby decreased, further adding to the resolution efficiency of this system. This device is therefore the most versatile and useful for estuarine studies. Not only can the general features

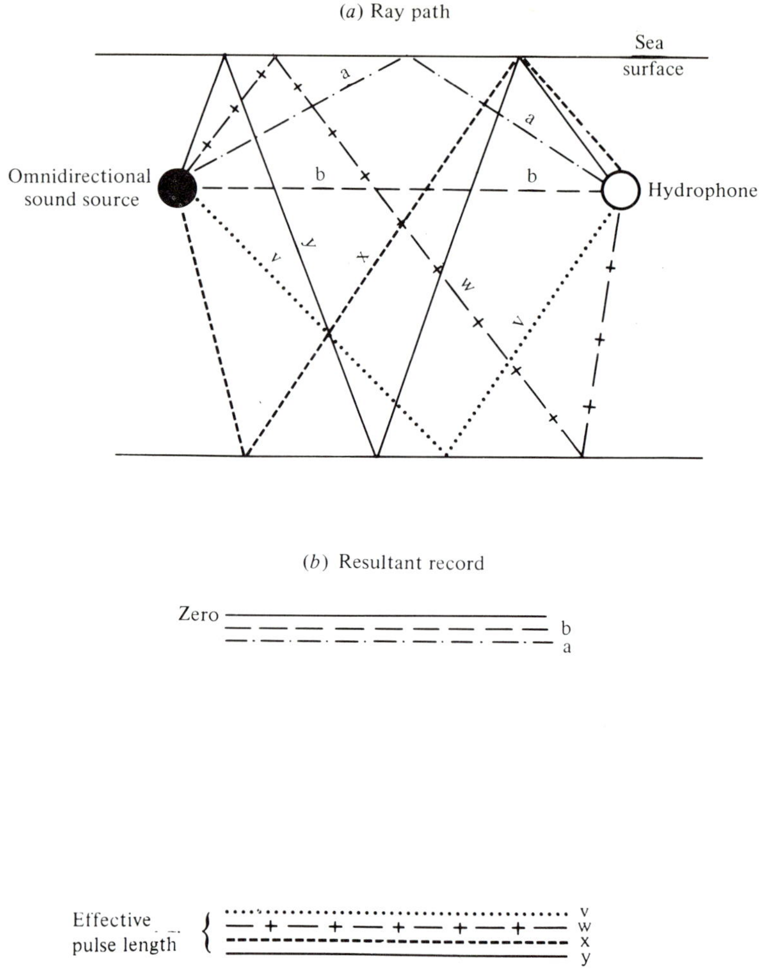

Fig. 4.8. *a*. Ray paths between an omnidirectional sound source and hydrophone; *b*. Resultant record showing how the various ray paths contribute to the effective pulse length.

of the sedimentary layers and the broad structural characteristics of the underlying bedrock be ascertained but, for example, details of the internal larger scale cross-bedding of sandbanks and sandwaves and other important depositional evidence can be picked out. Penetration, particularly in less consolidated fine grained sediments, can also be appreciable, sometimes in exceptional circumstances exceeding even 100 m.

Pinger. When resolution is all important the Pinger probe is the most useful piece of equipment. The sound sources used are similar to those mounted in side-scan sonar devices and are usually piezoelectric or magnetostrictive transducers. They are either hull mounted or set in a towfish to operate at fixed narrow-band frequencies, although some systems can have a variable operating frequency to suit the suspected geological conditions. The equipment is designed to emit a focused beam of sound, thereby allowing little or no energy to be transmitted to the surface. The O.R.E. 1036 system, for example, operates over a variable frequency range of 3.5–7.0 kHz with a variable beamwidth of 55°–30°. Pingers operating above 6 kHz can only effectively penetrate highly porous, low density silts and clays although this makes them particularly effective in estuaries where these grades of sediment occur so commonly. Resolution can be as little as 0.2 m with penetration of up to 50 m in the most ideal situation. However, penetrations of between 5 and 30 m are most likely whilst in high density sediments the device might operate with only a metre or so of penetration.

Receiving and recording

The returning reflected pressure pulses are, in the case of a Pinger, received by the same transducer switched automatically to the receive mode. Transducers are inherently narrow frequency band transmitters, wide band receivers, so some filtering is necessary to eliminate ambient noise and transmitter ringing. To receive reflected signals generated originally by a Sparker, Air-gun or Boomer it is necessary to stream behind the vessel a hydrophone which contains a variable number of piezoelectric or occasionally magnetostrictive elements mounted in a flexible, oil-filled tube. This refinement helps to achieve a certain degree of neutral buoyancy, good acoustic coupling and additionally aids in cutting down water noise whilst underway. The number of elements depends to some extent on whether resolution or penetration is the prime aim. Large numbers of elements are necessary for deep penetration studies. The spacing of the hydrophone elements is designed so that they receive efficiently the in-phase reflections from the bed and sub-surface interfaces but damp or cancel noise and pressure variations emanating from the ship and its independent motion.

The detected signals are fed from the hydrophone via amplifiers and sets of variable filters to graphic recorders which print the signal on dry or wet electrosensitive paper. A limitation of primary graphic recording is that the quality of the record is determined by the operator by subjective evaluation as the work proceeds. The quality can depend amongst other things on correct filter settings, a proper degree of signal amplification and on recording stylus sweep-speed. The latter is of especial importance when using Pinger equipment. Here, shallow penetration with a consequent small two-way pulse travel time necessitates fast stylus speeds so that small features within the sediments can be resolved and recorded and are not lost due to compression of the records. When deeper penetration is being attempted slower sweep speeds can be used. However, if the recorder has the facility, magnetic tape records can be taken and replayed in the laboratory onto a graphic recorder at a variety of filter, stylus sweep and other settings.

Continuous seismic profiling operation and interpretation

In order to achieve high resolution particularly with omni-directional sound sources such as sparkers and air-guns, it is necessary to tow the source and receiver close to the sea surface. However part of the emitted signal is reflected from the water surface undergoing a phase change and combining then with the downward directed part of the acoustic pressure pulse to lengthen the effective pulse length. Towing close to the sea surface cuts down the length of this unwanted surface reflected wave. It has been found that if the source and receiver are towed at a depth equal to a quarter of the wavelength of the emitted wave then signal strength is increased (Leenhardt, 1969). In operation this is difficult to achieve so towing close to the surface within one half wave length of the dominant emitted frequency is the practical ideal. Near-surface towing, however, is likely to pose problems concerning the reception of sea and ship generated noise so that trial and error is often necessary and the ship speed, towing depth and receiver/source separation may be altered to achieve 'low-noise' recording. Deep towing, although generally not applicable in most estuaries, is a technique useful in overcoming this problem.

Some of the acoustic energy on reflection from the sea-bed and sub-surface layers will reflect again from the water surface and return to the sea-bed or a sub-surface layer to be reflected again before being picked up by the receiver. This may happen continually or just occasionally but these secondary or multiple reflections make interpretation of the resulting records difficult at times (Fig. 4.9). The multiples generally mask the 'real' events received from sub-surface depths greater than the water depth. Although skilful use of filter settings and the adjustment of the relative positions of receivers and source, coupled where practicable with laboratory signal processing techniques, can partially overcome this problem, it remains a serious drawback when carrying out shallow water work, particularly where high levels of penetration are required.

A feature common where rugged surfaces occur within a sediment cover is the hyperbolic point source (Fig. 4.9). This is produced by a diffraction of the emitted signal from any surface which has a broken, uneven surface.

Commonly, in estuaries which have resulted from the flooding of the lower reaches of rivers, high amplitude C.S.P. reflections giving a characteristic

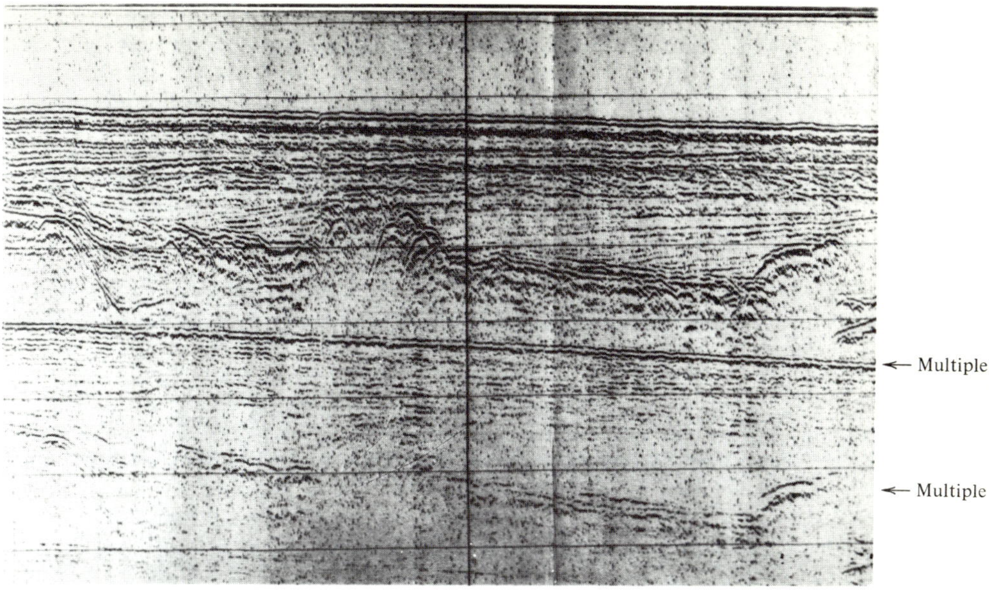

Fig. 4.9. CSP record showing multiple reflections occurring at twice the travel time of the first reflection. Note also the hyperbolic returns in the centre of the record from the relatively rugged surface of a buried 'Chenier ridge'.

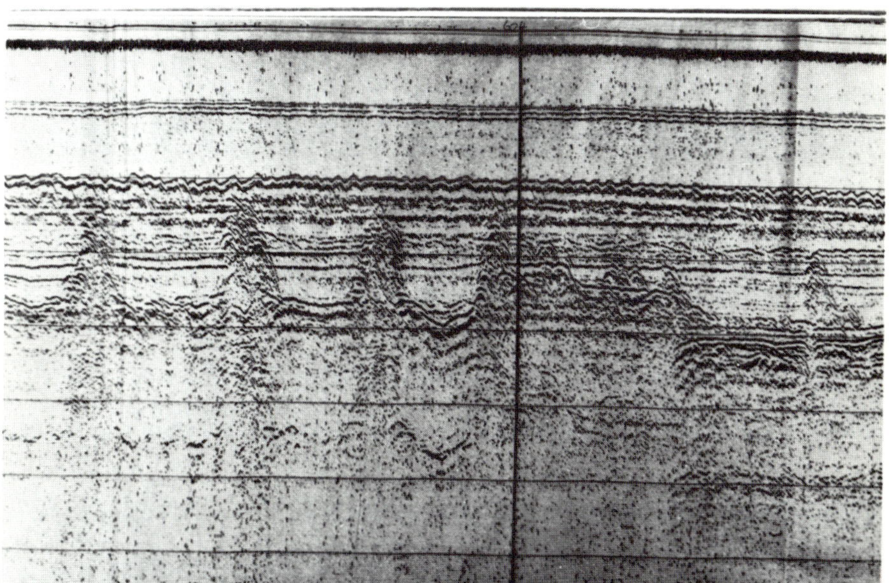

Fig. 4.10. CSP record showing 'bright spot' reflections from buried peaty deposits.

'bright spot' effect (Fig. 4.10) have been frequently observed. These have been interpreted as highly organic silts and clays often deposited originally on the river flood plain. More recently, reports of high pressure gas at shallow depth in some North Sea boreholes have drawn attention to the likelihood of gas pockets producing similar effects. Gas is likely to be a concomitant of the thick peat deposits one might expect in the post-fluvial estuarine environment. Another interpretation is that these sub-surface layers of high acoustic impedance, and where penetration is so often substantially reduced, are thick accumulations of shell and might represent buried 'Chenier ridges' or large off-shore shell banks. This has certainly proved to be the case in one instance in the Thames Estuary (Fig. 4.9).

Survey planning

Underlying the estuaries of today are to be found thick sequences of Pleistocene and Recent sediments overlying 'bedrock' of widely varying age: a series which can be elucidated by using continuous seismic profiling equipment. The use of this equipment certainly enables the estuarine investigator to obtain a continuous section of the underlying sediments and strata with a facility normally denied to his land-based colleague. He has to contend, however, with problems of expense in the purchase or hiring of equipment and boats, sometimes inclement weather conditions which generate high noise levels, and of equipment malfunction. When all things combine to co-operate, geological profiles up to 160 km in length or even more can be achieved in a 24-h period. The estuarine geologist has other problems to tackle. In the outer section of the estuary he may be able to run his survey traverses in any direction he chooses, thereby obtaining true-dip sections, strike sections and crossing other major features in the optimum way. Within the inner reaches the problems of shallow water over sandbanks, bars and tidal flats, which preclude safe navigation, have to be met. These can be partially offset by tackling the areas at periods of high water or by using shallow draught vessels or hovercraft. Very shallow water work can only be attempted by employing land based refraction techniques using gelignite explosions or the dropped weight impact system as sound sources. Further problems arise here and it is generally found necessary to resort to borehole data in the shallow water zone.

Heavy shipping traffic in the main channels between sandbanks, wrecks, buoys and other obstructions all prevent the ideal survey traverses being run. Once started, the geological features that are found will dictate the overall survey operations.

Before a survey begins it must be decided whether to aim for good resolution of buried features by using a Pinger probe, effecting a compromise by using the high resolution Boomer or aiming primarily for penetration by using a Sparker or Air-gun system. Sometimes it is possible to run more than one system at once, namely a Pinger and Boomer. Following this, a study of nearby land areas, geological maps, and borehole logs will provide some information as to lithology, thickness, strike and dip of the underlying rocks in addition to such details as the dominant trends of faults and fold axes. Recourse to both land maps and hydrographic charts might provide some clues as to possible former courses of rivers, subsequently drowned, whilst further information as to the likely position of off-shore sediment bodies might be obtained from the charts. All these factors will complicate the decision on where to run the survey lines and unless the investigator is very lucky he will have to effect some compromise.

C.S.P. survey examples

Information that might be gained concerning the sub-estuarine deposits using a variety of continuous seismic profiling methods might include any of the following.

Underlying any estuary, most of which will have been formed by flooding of the lower reaches of rivers by the Flandrian transgression of the last 18000 yr, will be 'basement' rocks. These might be of very variable age such as in the Severn Estuary, where rocks ranging in age from the Devonian/Carboniferous up to Miocene might be encountered, whilst in the Thames Estuary, the 'basement' is Cretaceous Chalk and Eocene marine and estuarine sands and silts. Within these basement rocks the dip of the individual units, their variation of thickness, and any major lateral facies changes might be discernible. Most rocks have a particular acoustic 'signature' that can be readily recognised by an experienced investigator so that if some traverses are started either close to a borehole site or from a well-exposed shore outcrop, the various beds can be traced off-shore and there matched with their acoustic signatures.

In addition, faults, their trend, throw and angle of hade, as well as any relationship that might exist between them and any areas of folding can be discerned. Fig. 4.11 shows a large monoclinal structure where the steep limb has in some places been faulted. This off-shore fold structure had been predicted by previous examination of the Bouguer gravity anomaly map for SE England, which showed a large negative anomaly in the area, and by the report of the thinning of several prominent Tertiary and recent deposits over its continuation on-shore in South Essex. The lines of faulting were found to be similar to the dominant fault trend in the Kent coalfield some 40 km away to the south-east.

In a similar way intrusive and extrusive igneous bodies such as sills and

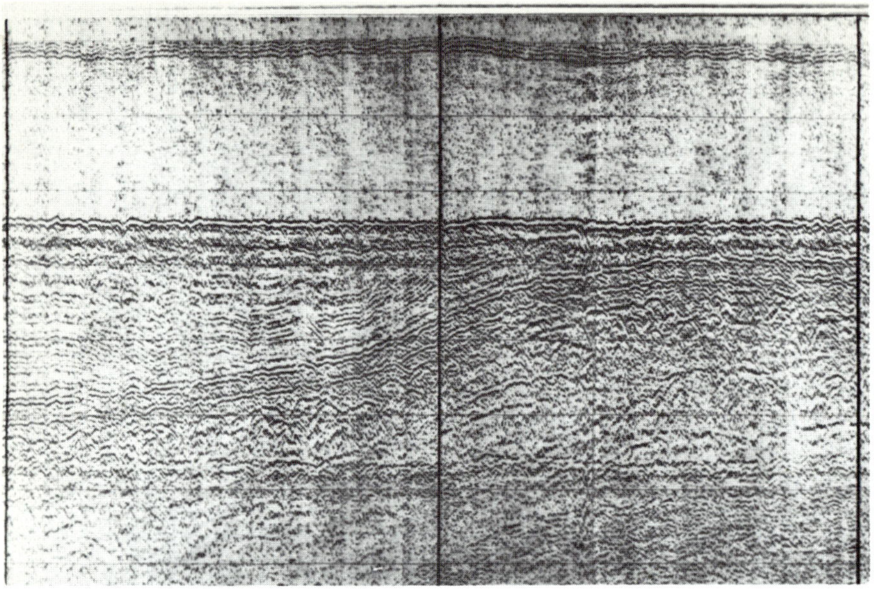

Fig. 4.11. CSP record of monocline, with Lower Tertiary clays overlying Cretaceous chalk.

dykes, volcanic cones and granite bosses will usually be plainly evident and even to some extent predictable.

From all this information a geological map can be drawn showing the outcrop pattern at 'basement' rockhead as well as a series of maps indicating depth to various sub-rockhead marker horizons. In addition, a map of the depth to the rockhead below a fixed datum, such as Ordnance Datum Newlyn, can be drawn as this is often of geomorphological interest indicating the course of Pre-Flandrian river valleys, their tributaries and other features.

Lying above this older series of rocks will be the Pleistocene and Holocene sediments, still, in many cases, retaining evidence of their mode and environment of formation. Features of glacial origin might become evident after extensive profiling and detailed mapping, although absolute confirmation would have to await borehole sampling in vital areas. Cutting through and partially destroying these features in once glaciated areas, or emplaced into the basement rocks in non-glaciated areas, will be evidence of river activity at times of lower sea-level. This will include terraces with their contained gravels and sands (Fig. 4.12). Occasionally, levees, point bars, ox-bow lakes, flood plain sediment, as well as channels containing thick deposits, will be evident. A map of the course of these rivers and their tributaries can be made, marking in such features, and especial note made of the relationship of such drainage systems with the regional strike and of fault and fold patterns. In the case of the former course of the River Thames, now lying buried under its estuary, it has been found that relative uplift during and since the Pleistocene period on several notable anticlines has caused the river to cut a path on the flanks and around the ends of these features.

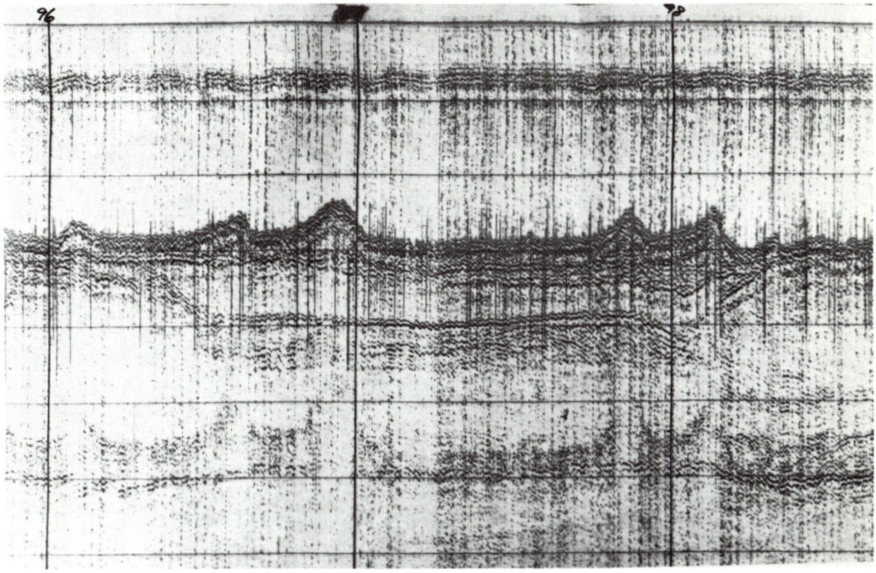

Fig. 4.12. CSP record showing a buried river valley. Note the sandwaves on the sea-bed.

Overlying all this will be evidence of the returning influence of the sea. In places, any or all of the previously mentioned features will be modified, often leaving a lag sand and gravel overlying the older Pleistocene and Holocene sediments. Sometimes this lag is seen to be banked against islands of earlier sediments or of 'basement' rocks as temporary beach deposits before being finally overwhelmed and buried under later deposits of the steadily rising, transgressive sea. Evidence of the order and form of these depositional processes can be seen as medium and large-scale cross-bedding and from these structures the final sediment transport direction can be inferred. Under tidal conditions and where sediment was in plentiful supply large sandwaves and sandbanks will have formed leaving evidence of their macrostructure discernible when using continuous seismic profiling methods (Fig. 4.13). Sometimes on the edge of inter-tidal flats, shell and coarse sand sorted by breaking waves will be banked into ridges which might then migrate across the flats under the influence of storm waves. These 'Chenier ridges' are sometimes found as fossil forms and Fig. 4.9 shows such a ridge, proved by boring, which in this case is some 10 km long, and some 35 m wide.

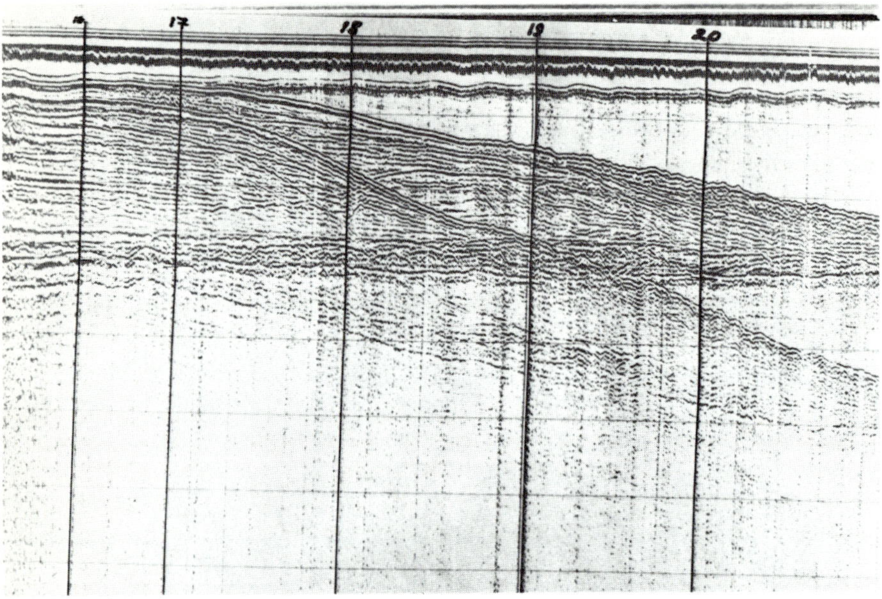

Fig. 4.13. CSP record showing the internal structure on the edge of a large sandbank. Note the multiple reflections.

Using the CSP data and other evidence it is therefore possible to work out the underlying structure and structural history and the sedimentological and palaeoenvironmental history of an estuary. With borehole data this framework can be given lithological, palaeontological and geochemical substance. In addition, the time frame for the various events can occasionally be substantiated using the C^{14} dating system where peat deposits, fossil wood and shell beds are encountered.

Industrial applications

A recent example has been the proposed Maplin Scheme for the reclamation of the Maplin and Foulness Sands in the Thames Estuary in order to build an airport, a seaport, and an associated industrial complex. The initial site investigation work involved a continuous seismic profiling survey over the sands at high water using a hovercraft towing a precision Boomer. Because of the shallow water, multiple reflections were a serious problem but sufficient information could be gleaned about certain very strong reflectors at depth under the sands, to build up an accurate picture of the underlying structure of these tidal flats. Borehole information gave both initial guidance to the survey and later refined the interpretation of this succession of recent sediments.

With the growth of the construction industry the need for sand and gravel has grown apace. Sources of land-based material have become seriously depleted in some areas so that operators have had to look off-shore for further supplies. Here vast quantities of sand, free from appreciable proportions of clay and shell contaminants, can be exploited, much of it readily accessible in estuarine areas and close to the major markets. Half a sandbank, some two hundred million cubic metres of sand was to be removed and pumped on to the Maplin Sands to effect its reclamation, and this without seriously affecting the tidal flow characteristics of the Thames Estuary. Gravel, not usually so readily available, however, has been found using both side-scan sonar and continuous seismic profiling techniques, particularly in those areas within estuaries suspected of once being the site of glacial moraines and outwash spreads. Also, the former sites of drowned river terraces and beaches have proved to be extremely productive areas of off-shore gravel.

With the advent of deep draught vessels, port authorities have found it necessary to investigate routes where channels might be dredged to below naturally maintained depths. This would involve, amongst other things, a complete sub-bottom profiling survey along all the projected routes in order to evaluate the likely sediment and rock conditions to be encountered. Routes involving the dredging of soft uncemented, recently-deposited sediments are likely to be favoured over those where hard rocks outcrop frequently. However, lightly cemented 'soft' rocks will stand better, free of slumping after dredging, than the very mobile sand and silt so often found cloaking the estuary floor.

By using both side-scan sonar and the sub-bottom profiler, a detailed picture of the recent sediments and the underlying rocks can be formed, a picture of research and commercial importance.

REFERENCES

Belderson, R. H., Kenyon, N. H., Stride, A. H. & Stubbs, A. R. (1972). *Sonographs of the Sea Floor. A picture atlas.* Elsevier Publishing Co., Amsterdam.

Chesterman, W. D., Clynick, P. R. & Stride, A. H. (1958). An acoustic aid to sea bed survey. *Acustica*, **8**, 285.

Flemming, B. W. (1976). Side-scan sonar: a practical guide. *Int. hydrographic Rev.* **LIII** (1), 65–91.

Kelland, N. C. (1972). Preparation of isometric records from high resolution side-scanning sonar data using the Bath University playback equipment. *Internal Report 72/73. Unit of Coastal Sedimentation, Taunton.*

Langhorne, D. N. (1973). A sandwave field in the outer Thames Estuary, Great Britain. *Mar. Geol.* **14**, 129–43.

Leenhardt, O. (1969). Analysis of continuous seismic profiles. *Int. hydrographic Rev.* **XLVI** (1), 51–80.

Taylor Smith, D. & Li, W. N. (1966). Echo-sounding and sea-floor sediments. *Mar. Geol.* **4**, 353–64.

GENERAL READING

Buchan, S., Dewes, F. C. D., McCann, D. M. & Taylor-Smith, D. (1966). *Measurement of the acoustic and geotechnical properties of marine sediment cores.* Marine Sci. Lab., Geol. Rept. 3.

Burton, A. N. (1976). *The use of geophysical methods in engineering geology: Part 1: Seismic Techniques.*

Chesterman, W. D. (1974). The Ocean floor. *Contemp. Phys.* **15**, 501–16.

Cole, B. F. (1965). Marine sediment attenuation and ocean-bottom reflected sound. *J. Acoust. Soc. Amer.* **38**, 291–7.

Dobson, M. (1969). The oblique asdic and its use in an investigation of a marine high-energy environment. *Sedimentology*, **13**, 105–22.

Hopkins, J. C. (1972). Tape recording of side-scanning sonar signals. *Int. hydrographic Rev.* **XLIX**, 59–70.

Ingham, A. E. (ed.) (1975). *Sea Surveying.* J. Wiley & Sons Ltd, Chichester.

Knott, S. T. & Hersey, J. B. (1956). Interpretation of high resolution echo-sound techniques and their use in bathymetry, marine geophysics and biology. *Deep-Sea Res.* **4**, 36–44.

Knott, S. T. & Bunce, E. T. (1968). Recent improvement in techniques of continuous seismic profiling. *Deep-Sea Res.* **15**, 633–63.

Leenhardt, O. (1971). Resolution in continuous seismic profiling. *Int. hydrographic Rev.* **XLVIII**, 71–83.

Leenhardt, O. (1974). Side-scanning sonar – a theoretical study. *Int. hydrographic Rev.* **LI**, 61–80.

Parasnis, D. S. (1962). *Principles of Applied Geophysics.* Methuen & Co. Ltd, London.

Sargent, G. E. G. (1968). Application of acoustics and ultrasonics to marine geology. *Ultrasonics*, 23–38.

Stride, A. H. (1961). Geological interpretation of asdic records. *Int. hydrographic Rev.* **XXXVIII**, 131–9.

Stubbs, A. R. (1963). Identification of patterns on asdic records. *Int. hydrographic Rev.* **XLI**, 53–68.

Wong, H. K. & Chesterman, W. D. (1968). Bottom backscattering near grazing incidence in shallow water. *J. acoust. Soc. Amer.* **44**, 1713–18.

5

Sediment sampling and analysis

A. T. BULLER AND J. McMANUS

In much the same way as the physical constraints of an estuary permit wide dynamic variability in different estuaries, the sediments in transit through and deposited within estuaries vary greatly from locality to locality. Since the deposited material forms the substrate upon and within which many of the estuarine organisms live and from which many draw their food supply, the sediment serves as an important factor in the estuarine ecosystem. The fine particulate material derived from the land or the sea and passing into the estuary may undergo substantial changes in chemical state, and may entrap pollutant materials. The geochemical gradients frequently detected in the uppermost zones of deposited muds and silts because of the mobilisation of certain elements and enrichment at the surface indicate the importance of this part of the sediment column. The sand and gravel-size sediments may be of economic significance as building materials. All sizes of material are potentially important in the siltation of navigable waterways and harbours.

The sediments which are so significant to many facets of estuarine studies are composed mainly of mineral particles derived from the breakdown of rocks of the drainage basin, local cliffs or the nearby sea-bed (Guilcher, 1967). The fragments vary greatly in composition, shape, and size and respond differently to movement of the water. It is the size of the bed particles which most closely reflects local water dynamics and also exerts a control on benthic organisms. The range of sizes present at one site determines the packing of the particles and the availability of pore spaces through which interstitial fluids may pass. In this way the sediments strongly influence the nature of both organic and chemical activity in the bed. Thus some form of investigation of the sediments is involved in virtually every estuarine study.

The aim of this contribution is to provide an outline of the principal methods of obtaining samples of sediment, the manner in which these materials may be processed in the laboratory to determine particle sizes, and to give some general statement on how sediment size analyses may be used to provide information about an estuary. The chapter is not intended to serve as a comprehensive manual for all sampling equipment and methods of grain-size determinations but rather to provide a guide to the main apparatus and analytical methods commonly employed in studies of estuarine sediments. Many laboratory manuals exist which deal with procedures, some excellently detailed, and the reader is referred to these if greater detail is required for any method outlined. Useful reference texts are Krumbein & Pettijohn (1938), Irani & Collis (1963), Müller (1967), Allen (1968), Carver (1971) and Folk (1974), which may be profitably supplemented from British Standard 1377 (1975) and publications of the American Society for Testing Materials. Sampling apparatus is reviewed in Hough (1939), Shepard (1973) and by Kelland (1975).

Sampling methods

As the physical nature of the estuary varies, so must the techniques of sediment sampling employed. In many coastal plain-type estuaries extensive areas of low-relief tidal flats may be covered with shallow water for part of the tide, but are exposed for much of the tidal cycle. These may be sampled either from a ship or directly by hand at low tide. Bottom sampling by means of grabs or corers is relatively straightforward in low currents, but may be rendered impossible during the strongest currents. The deeper fjord estuaries often characterised by low speed currents may be sampled at almost any time using grabs or corers but, since substantial depths of water are involved, simple hauling of lines may be impracticable and winches are required.

The catalogue of instruments which have been developed for sampling bottom sediments under water is lengthy. Two principal groups of apparatus exist: grabs and corers. In most cases they have been designed for deep sea oceanographic or Continental Shelf surveys, but many have undergone miniaturisation so that they are suitable for deployment from small boats without sophisticated winching facilities.

Grabs

The grabs most commonly used fall into three main groups. All function through the lowering to the bed of a scoop or bucket of specialised shape. Impact with the bed releases a restraining mechanism, and sampling occurs with withdrawal.

The Pettersson grab is one of the longest surviving forms. It comprises a pair of quarter cylinder buckets hinged so that they may be held apart as the grab is lowered to the bed, released on contact, and lifted from the hinge line. This clam type of grab achieves samples to a depth determined by the radius of the cylinder. Sediment may escape if the jaws are jammed open by gravel or shells during lifting.

The van Veen sampler (van Veen, 1936) is rather similar in operation, but the jaws of the grab are held together firmly during withdrawal, for the lifting cable is attached to the ends of the arms which protrude past the axis and exert a strong clamping action. This form of grab normally has inspection hatches in the flat side of the sampler bucket, which enable inspection of sediment surface structures or withdrawal of small cores before release of the sample.

The addition of teeth to the jaws (Fig. 5.1) enables the grab to penetrate and cut more readily into closely packed sandy bottoms (Lees, Buller & Scott, 1969).

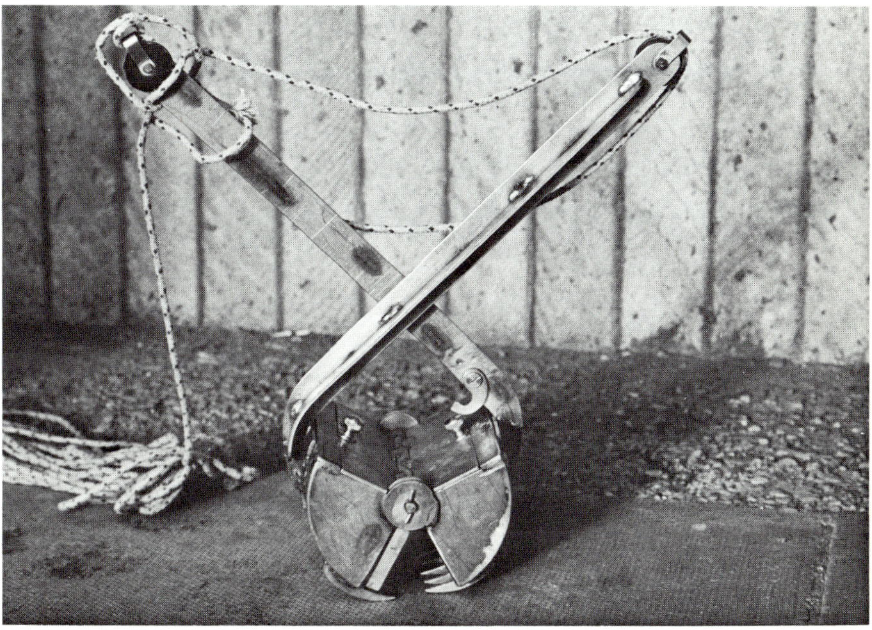

Fig. 5.1. Modified van Veen grab, showing added teeth and the observation ports.

The third type of apparatus, the Shipek grab (Fig. 5.2) is widely used in shallow marine investigations. It consists of a semi-cylindrical bucket which is lowered to the bed open side downwards. As it touches the bottom a powerful spring is activated, causing the bucket to rotate into the bed. The spring holds the sampler tightly closed during retraction.

All three grab samplers are suitable over the entire range of sediment sizes. However, the Pettersson may leak more readily than the van Veen, with attendant loss of fine particles. The Shipek is undoubtedly as effective for fine

Fig. 5.2. Shipek grab. (Photograph: courtesy Hydro Products Ltd.)

bed materials, but is far superior for sampling gravelly deposits. However, the very powerful spring trigger on this form of grab must be set before it is lowered to the bed. This is not easily achieved in small launches and once primed premature or accidental release of the trip mechanism may be a hazard.

Corers

If information is required about vertical changes in the sediment beneath the surface the often greatly disturbed grab samples may not be adequate, and the use of coring devices becomes necessary.

The simplest gravity corer consists of a metal tube which has weights and stabilising fins at the upper end. The corer penetrates the bed after free fall through the water. The sample passes through a core cutter which has a diameter slightly less than the corer tube above. This is lined with a plastic core liner tube. The decrease in diameter reduces frictional resistance to penetration exerted by the tube walls and reduces the compression and distortion which would otherwise occur at the core margins. The core is withdrawn in its liner and may be stored for later analysis. Then both the core and liner are cut lengthwise and sediment samples taken serially as required. Short barrelled gravity corers can produce cores up to 0.5 m in length in sands, but generally sands are densely packed and are difficult to sample. Longer samples can be obtained in silty and muddy deposits. The use of a core retaining spring or core catcher on the cutting nozzle to aid in retention of the sediment during withdrawal is often advantageous in the finer sediments, but is less valuable in sandy deposits.

The penetration achieved is partly related to the length of the corer tube, the depth of water for free fall, the number of weights attached, and the nature of the bed sediments. The number of lead weights which can safely be attached is restricted in small boats lacking winching facilities. However, useful cores may be obtained in most fine sediments in estuaries with minimal weights.

Piston corers, such as the Kullenberg corer (Kullenberg, 1947) are widely used in deep sea surveying. Impact with the sea-bed releases a piston which produces a vacuum within the core liner. This helps to reduce internal friction and compression of the core, resulting in improved penetration. The use of piston corers is not excessively complex and has the great advantage that it yields virtually undisturbed cores of sediment which may be several metres in length in suitable substrates. However, since fairly heavy weights are involved it is essential to have winching facilities available for use with this equipment.

Modifications of the piston corer by Silverman & Whaley (1952) and later by Ginsburg & Lloyd (1956) have permitted it to be used with some success in shallow waters of the near-shore coastal zone and estuaries.

When free-falling or piston corers encounter rockhead or gravelly deposits, buckling of the cutting edge or even bending of the core barrel may occur, so care should be taken to avoid the use of these devices in situations where rock is likely to be met.

In the last decade increasing use has been made of corers whose penetration is achieved through vibration techniques. The vibrocorer is capable of obtaining undisturbed continuous sediment cores of several metres length. The equipment, which incorporates a frame and vibrator mechanism around the core barrel, is lowered to the bed before being activated. Excellent results can be achieved in sands which are incapable of being sampled by other techniques. However the instrument is generally heavy and requires a substantial support vessel for operation. A detailed assessment of vibrocorer performance was made by Kirby (1972).

A second form of apparatus used both on land and under water is the box-corer. This is essentially for studies in which the internal structure or fabric of the sediment is under examination. An open steel box is driven vertically into the sediment until the top rests on the surface, and then a blade cuts through the sediment and covers the open end of the box. The box is lifted clear and the monolith of sediment may be examined in undisturbed form. The box-corer boxes commonly used range up to 50 cm in depth, 30 cm in length and 20 cm width. Heavy weights may be added depending on the nature of the substrate. On the land investigators can excavate the surrounding sediment to release the encased samples. Although heavy gauge steel is normally used for these boxes, 4 l oil cans from which the base has been removed can be used for the same purpose. Extrusion of the sample from the box is best achieved by electrolytic methods, using a 36 or 48 volt DC current to drive water to the surrounding can walls to lubricate the margins of the sediment monolith and so release it intact for examination. Samples collected by box-coring methods are heavy and there are practical restrictions on the number which can be collected.

The box-corer has been successfully modified for use at sea by Reineck (1963) and has been used to obtain excellent records of estuarine sediment sequences and Wadden Zee deposits.

The size of sample obtained by box-coring methods enables the organisms to be detected in place within the sediments and the resultant bioturbational features examined or preserved using resin impregnation techniques.

Other techniques

In his review of bottom sampling equipment Hough (1939) noted that although towed bucket or dredge samplers may retain some sediment, notably pebbles and occasionally lumps of fine sediment, none gave consistent recovery of sediment from mixed or sandy bottoms. Dredge samplers are designed principally to recover rock fragments from the bed. Most consist of a heavy metal frontal frame behind which is towed a bag into which materials dislodged by the dredge bar are carried. Normally only pebbles, rock fragments, organic remains and occasional mud lumps are retained, for unconsolidated sand, silt and clay are washed out as the dredge is towed by the survey vessel.

Divers using SCUBA systems may provide valuable observational data on sediment surface structures, distributions of organisms or systematic local variations of particle size and may undertake bottom sampling programmes. In many estuaries underwater visibility is severely restricted by high concentrations of fine particles in suspension, and in many places manual sampling by diver may be carried out only during slack water. Although there are clear water estuaries with generally low currents they are not common and consequently manual collection of bottom samples by diver is not a widely used technique.

Sampling design

Sediment sampling from a ship is often carried out on a grid pattern. This bears no relationship to the scale of the sediment distribution pattern and may yield a very biased and unrealistic result. It is better to carry out a side-scan sonar survey initially and then determine the number and location of samples accordingly.

Sampling sediment on exposed tidal flats and banks poses problems which tend to become overlooked where remote sampling is carried out. How many samples should be taken, at what spacing, how large should each sample be, should they come only from the surface or to what depth should cutting be permitted? The frequency of sampling and size of sample needed depend very much on the requirements of the investigator. For a simple size distribution pattern all significant sediment types should be sampled as discrete entities and at intervals defined by the size of the area to be examined and by the local complexities. In studies seeking to interrelate physical controls and the resultant sediment deposition the sediment needs to be undisturbed, for a single layer will have been deposited by a single process, so that the entire sample may need to be taken from that single lamina. A sample of one kilogram is normally more than adequate to permit the entire range of sediment analyses to be performed, and in fine sediments lesser quantities are needed (Table 5.1).

Table 5.1. *Maximum weight of material to be retained on each sieve on completion of sieving*

B.S. sieve mesh (mm)	Maximum weight (kg)	Diameter (mm)
20	2.0	300
14	1.5	300
10	1.0	300
6.3	0.75	300
5	0.5	300
3.35	0.3	200
2.0	0.200	200
1.18	0.100	200
0.600	0.075	200
0.425	0.075	200
0.300	0.050	200
0.212	0.050	200
0.150	0.040	200
0.063	0.025	200

Vertical profiles may be obtained by trenching to establish stratigraphic sequences, patterns of local micro-environmental succession, variations of sediment type or changes of pollutant concentrations with depth. Simple spade or trowel cutting is all that is needed, with the sediment samples obtained being placed in bags or bottles for transport to the laboratory.

Sample preparation

Sediment samples rarely reach the laboratory singly, and it is most important that soon after their arrival a check is carried out. The sample bags, bottles or cores should be laid out sequentially and the numbering and labelling checked. External labelling may have become obliterated during transport, and systematic curation on arrival may enable identification through checking the internal label before that too becomes obliterated or misplaced.

The extent and type of sample preparation will depend on the type of sediment with which one is dealing and the results that are required. The first decision will be whether the sediment will need to be sieved wet or simply washed and dried before sieving. If there is more than a few percent of clay in the sample then a hard crust will form on the surface of the sample after drying. Though this can be broken down with a pestle in a mortar, it is not to be recommended. Consequently wet sieving at 63 μm may be necessary and the resulting fine suspension retained either for pipette analysis or for measurement of this percentage of the total weight in the fine fraction. The coarse fraction can then be washed and dried for grain size analysis.

Should the sediment be a mud, wet sieving may still be necessary to deal with the small proportion in the sand fraction. The decision then is whether to analyse the suspension of fine sediment as it is, or after dispersion. If a complete grain size analysis is not required it may only be necessary to determine the proportions of clay, silt, sand and gravel for a quantitative description of the sediment type.

When one is dealing with the sand or gravel-size fraction one must additionally decide after washing and drying whether to analyse the sample with the incorporated shell material, or whether to remove the shell. The sample may need to be sieved both before and after removal of the shell.

Clay particles are abundant in many estuarine sediments, and the behaviour of these very fine grains is frequently determined by the identities, concentrations, and valencies of salts in the water. In some instances they exist as discrete particles, but in salty estuarine waters they cluster together to form flocs of a very much larger aggregate diameter than the component particles. Removal of the salts from the interstitial water modifies the micro-physico-chemical environment surrounding the clays and immediately destroys the delicate balance achieved in nature. If the salt is removed the subsequent size analysis may indicate the sizes of the individual elements but will probably fail to indicate the dynamic characteristics of the floc which existed in the real environment. Consequently there is a somewhat intractable problem associated with size analysis of fine particles, and the preparation procedure prior to analysis may largely determine the actual results obtained. The removal of the salts is still considered to be a normal prerequisite to the analysis although the comments just made must be borne in mind when attempting to interpret the results.

With the exception of samples collected primarily for purposes such as certain forms of soil mechanics testing or the investigation of micro-organisms between the sediment particles, etc., most estuarine sediments require to be dried before mechanical or chemical analysis is started. However, the drying of sediment which may contain substantial quantities of salt in the intergranular pore water demands some preliminary preparation. If the sediment with interstitial saline waters were to be dried as received in the laboratory a white crust of salt deposit would form on the surface and act as a cement binding together what may have been originally separate non-cohesive particles. In consequence the water-soluble salts must be removed.

Washing

If chemical examination of the sediment particles is to be undertaken total removal of the water salts may be required and the best method adopted is that of dialysis. The sample is placed in a dialyser bag which is in turn suspended in a container of distilled water, or in a circulating stream of distilled or deionised water. The salts are then removed by osmosis. When flowing water is used many samples may be simultaneously dialysed in a process usually extending over several days.

For sand or gravelly samples to be subjected to mechanical analysis for grain size determination the total removal of the salts is not vital. In most instances a large proportion of the salt may be removed by repeatedly stirring the sample in beakers of tap water, allowing it to settle and decanting the clear liquid. A final washing in distilled or deionised water removes low concentrations of salts. A useful figure to bear in mind here is that three washings in one litre of water are needed to remove the salt from about 200 g of sediment, with thorough stirring. The final wash should be left overnight. Care must be taken that fine particles are not lost during decantation.

Drying

The sediments are placed in open evaporating basins and inserted in a thermostatically controlled oven. The temperature and duration of the drying are largely dependent upon the size of the particles involved. Sands and gravels are dried for several hours at temperatures of 105–110 °C. When there is a significant proportion of clays in the sample, caking may arise. In addition structural damage may be inflicted upon certain clay minerals as water is driven from the lattice structure, so drying must take place at a lower temperature (50–60 °C) and is therefore a longer procedure. Any partial cementation from residual salts or caking of fine particles may be destroyed by gentle crushing between the fingers. Examination with a hand lens will reveal the presence of any further aggregates which may be eliminated by means of a wooden rolling pin, or a rubber pestle in a mortar. When clays form more than a few per cent of the sediment, drying should not be attempted until the fines have been removed by wet-sieving.

Sub-sampling

In order to obtain a representative sample most scientists collect more sediment than is required for simple grain-size analysis, and once washed and dried a truly representative sub-sample must be obtained. During each preliminary exercise some sorting by size occurs and it is important that the original sample should be rehomogenised before sub-sampling. This may be achieved by shaking the entire sample in a closed container.

When sufficient material is available the simplest method of splitting or subdividing the sample is by the process of quartering. The entire sample is poured on to a suitable surface in the form of a cone. The cone is then divided into four quarters. Two of the quarters are recombined, and a fresh cone formed, from which by further quartering the samples for analysis and storage in reference collections are obtained. This simple inexpensive method is inexact, but although subject to operator bias achieves a fairly representative sub-division.

In order to eliminate the potential subjective element, mechanical splitters are used in most laboratories. The chute splitter consists of a V-profile trough which is sub-divided into equal chambers feeding alternately into receptacles on either side of the axis of the trough. The sample is poured into the splitter and halved repeatedly until the desired quantity of sample is obtained. Some bias may be introduced by pouring more material at one end of the trough than at the other.

A third technique is to feed the sediment at a uniform rate through an orifice leading to a rotating ring of uniform containers. By this method many sub-samples may be obtained in a single process, but some limitation is imposed by the orifice aperture. The spinning splitter has a high sampling efficiency and eliminates operator bias.

Shell content

Shell fragments form an integral part of many estuarine sediments. These may be derived from and include the local fauna or have travelled considerable distances. Near the centres of high population density the shell content of the sediment may be high. The shell particles change in shape and size after death, and the valves and smaller fragments migrate in response to currents and waves (Wilson, 1967). Thus sediments at places remote from the present colonies of shell-forming organisms may incorporate comminuted shell debris.

Initially it is often sensible to remove the live shells as they are part of the mobile sediment. Since most shell particles are of platy or flaky nature they respond readily to water motion and the shell fragments ultimately come to rest in positions where their stability requirements bring them into sediments of very different grain size characteristics.

In order to assess the importance of shell material in the estuarine sediments, particularly the fine gravels and sands, it may be necessary, after a first phase of sieving, to immerse the sediment in dilute hydrochloric acid to digest the calcium carbonate mineral particles in the sediment. Comparison of the two analyses reveals the differences between the size distributions of the shell and mineral fractions.

The shells may also be used to indicate transport paths of sediments in the estuary, for the finding of identifiable particles of an organism to landward or seaward of the limits of its present habitat may signify net movement in that direction.

Grain size analysis

Before the investigator turns to the methods of size determination it is important to decide just what information is required. Size of particles is variably defined, but is usually taken as a measure of the dimensions which best describes the particle. Spheres may be characterised by their diameter, a cube by the edge length, a flake by its projection area, so the size is at least partly a function of particle shape. Most methods of determination are partially dependent upon shape, or assume that all particles approximate to spheres in the calculations used to determine size distributions.

If some general statement concerning the relative coarseness or fineness of sediments is required then the normal sieving techniques may be most appropriate. However, the important measure for potential cation exchange in clay particles may be the projection area. For each purpose there is usually a well-defined method available. In general for each range of grain sizes appropriate techniques exist, and the different methods are indicated on Fig. 5.3.

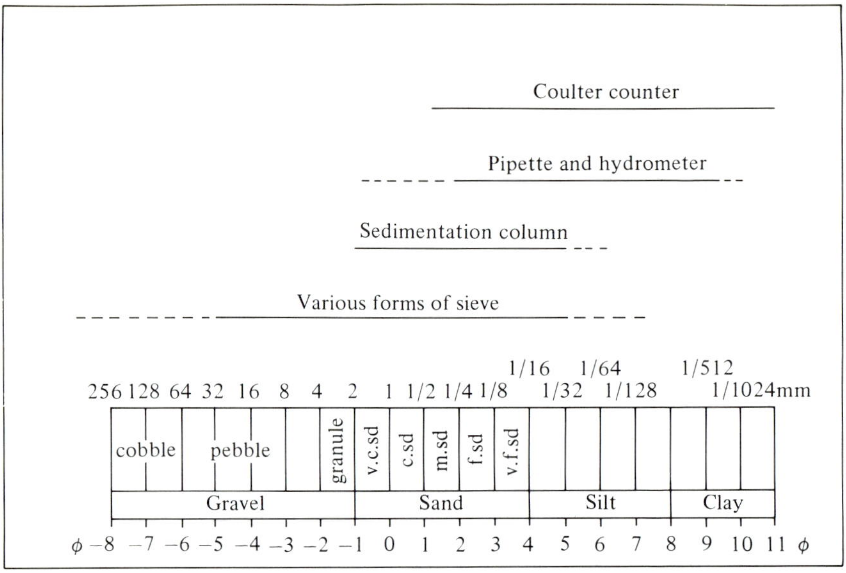

Fig. 5.3. Wentworth grain-size classification, together with the range of various analysis techniques.

Mechanical analysis

Dry sieving

The simplest and most widely used method of grain size analysis, dry sieving is appropriately used in the size range 0.063 mm to 250 mm. Mesh sizes are available down to 0.002 mm, but sieves with these apertures are expensive, easily damaged, and in practice are rarely used in estuarine sedimentology.

The coarsest sieves have strong wire or plates between the openings, those used for the sands are of fine wires arranged in a square mesh, and the number of holes per inch gives the mesh number. The operative size of each sieve should be checked using a microscope before use.

Preliminary preparation requires drying, disaggregation and if necessary removal and weighing of the silt and clay fraction. The weight of the sample to be used depends upon the average grain size of the sample. For analysis of gravels up to 10 kg is normally needed, for coarse sands 100–250 g is sufficient, but for finer sands 10–100 g is considered adequate. Sieves are normally arranged so that the meshes become progressively finer downwards at $\frac{1}{4}$ phi intervals. The sample should be shaken for a period of 15 min.

The material retained on the coarsest mesh sieve is now emptied on to a sheet of paper, and the sieve tapped gently diagonally to the mesh to dislodge trapped grains. Further gentle knocking on the bench will loosen more grains and the mesh should be cleaned with the appropriate sieve brush (wire for coarse meshes or camel-hair for fine apertures) in order that particles which have become firmly lodged may be removed. All sediment obtained is weighed to 0.01 g (or to 0.001 g for small residues). This process is repeated for each sieve, and the weight of any residue reaching the lowermost pan determined.

It is important to examine the individual sieve residues to detect the presence of aggregates. If these are present steps should be taken to disaggregate the sample using a mortar and rubber pestle and sieving recommenced once this had been completed. Misleading, invalid results might be obtained without the use of this form of checking.

Failure to clean the sieves results in a sharp reduction in the total aperture area available for grains to pass through, and a fall in the reproducibility of the results. A similar form of inaccuracy is introduced if the sieves are overloaded. This not only restricts the passages through which particles can progress to finer meshes, but may lead to distortion of the meshes themselves. The load per sieve should be at most 4–6 grains thick (McManus, 1965). An indication of the maximum recommended sieve loadings is given in Table 5.1 (see B.S. 1377 (1975)).

Increasing the sieving time will permit more sediment to pass down the column but, as indicated by Folk (1974), the optimum compromise time used by most sedimentologists is 15 min; this duration leads to acceptable reproducibility of analysis from one worker to another. There is a further potential variation in the style of shaker used, but, in practice, this is generally negligible.

A form for recording sieving analysis data which enables not only the raw weights and percentages retained on each sieve fraction to be seen, but also the values of any given grain size grouping to be shown and readily assimilated, is presented in Table 5.2.

Wet sieving

No single method of grain-size determination is adequate throughout the entire range of particle size encountered in estuaries and so, in order to obtain a complete grain size analysis, it is often necessary to use two methods in combination. Whilst sands and gravels are adequately dealt with by sieving, the silts and clays generally require analysis by means of pipette or hydrometer techniques. In order that the two sets of procedures may be followed, the fines must be separated from the sands. To separate them by dry sieving would be far from simple: drying of the sample may cake the fine particles and render disaggregation difficult; dry sieving to remove the fines is usually unsatisfactory because many fine particles adhere to the mesh surfaces and fail to pass through their appropriate niche in the sieve nest. In consequence most

Table 5.2. *Forms for recording sieving analysis data*

Sieve number	φ values	Approx. value (mm)	Approx. value (µm)	Weight retained (g)[a]	Weight/ Wentworth classification	Weight (%)	Weight %/ Wentworth classification	Cumulative (%)
				B.S. sieve analysis				
8	−1.04	2	2000		PG		PG	
10	−0.75				VCS		VCS	
12	−0.49				Sum		Sum	
14	−0.27							
16	−0.00	1	1000					
18	0.23				CS		CS	
22	0.52				Sum		Sum	
25	0.74							
30	1.00	$\frac{1}{2}$	500					
			Tay Estuary sediments programme					
36	1.24				MS		MS	
44	1.50				Sum		Sum	
52	1.76							
60	1.99	$\frac{1}{4}$	250					
72	2.24				FS		FS	
85	2.49				Sum		Sum	
100	2.72							
120	3.01	$\frac{1}{8}$	125					
150	3.26				VFS		VFS	
170	3.49				Sum		Sum	
200	3.71							
240	3.91	$\frac{1}{16}$	62					
Pan (loss)					St/Cl		St/Cl	
Total								

Original weight ....

Sieving loss ....

[a] Recombine fractions to full φ units and store

Weight % of $CaCO_3$ in sand

	Dry weight (data from column 6)	Dry weight after HCl treatment	Difference i.e. wt of $CaCO_3$	Wt loss of $CaCO_3$ as % of total	Wt % of $CaCO_3$	Cumulative wt % of $CaCO_3$
VCS						
CS						
MS						
FS						
VFS						
Total						

PG, pebbles and granules (including shells);
VCS, very coarse sand; CS, coarse sand;
MS, medium sand; FS, fine sand; VFS, very fine sand.
$\varphi = -\log_2$ (diameter in mm).

workers use wet sieving for successful separation. This is a slow process, which demands development of a skill by the operator, but it yields reproducible results.

No simple method enables the total fines content of a sediment to be directly measured. One can work from a known weight of material emplaced in a sieve nest and wash out and often lose the fines, thereafter determining total silt and clay by difference between weights of dry sediment before and after the exercise. Matched pairs of sub-samples can be taken and total analysis performed on one of the pair and only partial analysis on the other in order to retain the fine sediments for detailed examination. Alternatively the total weight of fine material washed through the sieve can be determined by drying a representative portion of the total suspension.

The procedure advocated in the British Standard 1377 requires that the sample shall be oven dried at 105–110 °C and weighed to 0.1% of its weight before being immersed in water containing a dispersant such as sodium hexametaphosphate (2 g/l). After an hour during which the sample is periodically agitated, and so progressively dispersed, increments are washed through the 2 mm sieve, which has been nested into the 63 μm mesh sieve. Gradually more and more of the sample is washed by flowing water so that the fine particles are separated from the sands, and are permitted to run away with the water. When the outflowing water is clear the sediment is removed from the sieves, oven dried at 105–110 °C, passed through the sieve nests as in dry sieving above, and each fraction is weighed in turn.

The total fines content is thus determined by the difference in weight between the sediment sample at the outset and the sum of weights retained on the individual sieves. This method of analysis, in which no more than 105 g may be held on the 63 μm mesh at the start of the experiment, leads to loss of the fine fraction and also requires very careful dispersion of the oven dried fine particles.

Since clay minerals become partially reconstituted at temperatures above 70 °C many researchers object to the initial drying required in this process, and also to the loss of sediment which could be used for size determinations by other techniques.

Perhaps the best alternative analysis requires subdividing the sample into identical fractions. The first portion is wet sieved, and the sediment passing the 63 μm mesh retained. Both the sand and mud fractions are dried and weighed in order that the relative proportions of the two parts may be established. Wet sieving of the sub-sample, retention, drying, and weighing of the sand fraction enables the weight of mud in the collected fraction passing 63 μm to be determined indirectly. This fine material is now available for further analysis.

A more rapid method may be used when more than 40% of fines are present. Duplicate sub-samples are weighed moist, one is dried and by subtraction the weight of moisture determined. The second subsample is wet-sieved and the retained sand dried and weighed. Knowing the total moist weight, the weights of the water, and the sand content, the weight of the fines may be determined. Again the mud is available for further analysis.

Pipette analysis

Many sedimentologists refer to the sieving methods of grain size analysis determination as 'direct' methods and the sedimentation methods using either pipette or hydrometer techniques as 'indirect'. Neither has the absolute truth for sieving yields results dependent upon the shape of the particles as well as the size, while the wet methods are based on Stokes' Law. Within the framework of Stokes' Law many interacting variables, such as particle shape, concentration, density and the efficiency of the dispersant action, influence the manner in which particles settle in a column of standing water. These additional potential sources of error render the wet methods less than totally satisfactory. Despite the drawbacks, and in the absence of better inexpensive methods, pipette analysis is normally applied to sediments possessing over 10% of fines in order to obtain detailed knowledge of the size distribution. It is possible to analyse the mud in its flocculated state providing the total concentration and the ionic concentration of the water are above certain levels (Pierce & Williams, 1966). However it is not possible to tell whether the sediment is in the state that it was in the estuary even though, with care, reasonably reproducible results can be obtained between sub-samples.

Dispersion of samples

Fine particles transported into an estuary may remain as discrete individuals or may become aggregated together to form flocs. Upon deposition the fine particles are brought into close proximity to similar particles and a form of ionic bonding or cohesion results. Attempts to analyse the size of the smallest elements of such a sediment, whether they reached the depositional area individually, in aggregates, or as flocs require that the interparticle cohesion shall be eliminated, and that all particles are separate from their neighbours. This separation is achieved by the use of peptisers or dispersants (as opposed to coagulants which may stimulate flocculation).

The most commonly used dispersant of many available is sodium hexametaphosphate, comprising 33 g of sodium-hexa-metaphosphate and 7 g of sodium carbonate dissolved in distilled water to give one litre of solution. The concentration of dispersant used by the authors is 50 ml of sodium hexametaphosphate *solution* to 1000 ml of distilled water. Throughout the analysis all wash bottles, etc., are filled with this mixture.

Sedimentation procedure

About 20 g of the material passing the 63 μm mesh during wet-sieving is thoroughly dispersed, mixed in one litre of water and allowed to settle. The large particles settle more rapidly than finer ones and so by repeatedly drawing off a small volume of suspension from the same position a record of the settling of the suspension past that point is obtained. The fluid withdrawn is evaporated and weighed, to indicate the quantity of suspended sediment remaining after a known interval. Careful selection of sampling times and depths permits the weights to be related to grain sizes and a curve of distribution to be constructed. Several analyses may be run simultaneously, thereby economising on time.

Opinion ranges widely on the sample size from 4 g to 35 g per 1000 ml. The authors recommend the use of 10 g and certainly no more than 20 g, for the greater the number of particles involved the greater the likelihood of flocculation in the sedimentation tube or of particles interfering with each other as they settle. Even in 1% concentrations the displacement of water creates upward moving columns of the suspension leading to retarded settling. This is known as settling convection (Kuenen, 1968). For flocculated suspensions, Pierce & Williams (1966) suggest that no interaction occurs when the concentration is less than 9%. The dispersed sample is placed in a measuring cylinder and dispersant solution is added to bring the total to 1000 ml. The sample and solution are stirred vigorously either by end-over-end mixing or by means of a manually operated stirring device which travels from bottom to top of the column of fluid. This ensures that fine sediment is uniformly distributed throughout the column, which should be maintained at 20 °C.

As stirring ceases a timer is started and a pipette is inserted into the standing fluid so that after 58 s, 20 ml of fluid is withdrawn from a depth of 20 cm below the surface. The fluid is released into a 50 ml beaker and 20 ml of distilled water used to wash any stranded particles from inside the pipette and into the same beaker. The beaker and cover glass are placed in an oven at 100 °C for at least 24 hr.

Successive 20-ml samples are withdrawn at intervals defined by the time taken for particles of specific sizes to settle through a known column height. In Table 5.3 are presented the times taken for particles of given diameter to fall through 20 cm, 10 cm, and 5 cm respectively as these are depths from which the 20 ml samples are normally taken.

Table 5.3. *Settling time for 20 °C and* S.G. *2.65*

Phi	mm	µm		depth (cm)	h	min	s
4.0	0.063	63		20			58
4.5	—	—		20		1	56
5.0	0.0213	31.2		10		1	56
5.5	—	—		10		3	52
6.0	0.0156	15.6	silt	10		7	42
6.5	—	—		10		15	
7.0	0.0078	7.8		10		31	
7.5	—	—		10	1	1	
8.0	0.0039	3.9		10	2	3	
8.5	—	—		10	4	5	
9.0	0.00195	1.95		10	8	10	
9.5	—	—	clay	10	16	21	
10.0	0.00098	0.98		10	32	42	
10.5	—	—		5	32	42	
11.0	0.00049	0.49		5	65	25	

After Krumbein & Pettijohn (1938).

During sampling the pipette must be gently inserted into the fluid lest local turbulence be created and sedimentation impeded. According to B.S. 1377 the pipette should be lowered over a period of 10 s duration and similarly withdrawal should also take 10 s. A standardised vacuum device for taking the sample can be set up using a filter pump attached to a water tap with a reservoir and a reduction system.

Successive 50-ml beakers are dried and the weight of the deposit determined to the nearest 0.001 g following cooling in a desiccator. The deposit will contain the sediment finer than the given grain size and also some dispersant. One-fiftieth of the dispersant weight used must be subtracted from the total weight of retained deposit in order to calculate the weight of sediment of grain size finer than the nominal diameter concerned.

Folk (1974) suggests the introduction of an initial sampling after 20 s have elapsed. This reflects the entire grain population before effective settling, i.e. starting at 63 μm, the coarsest material passing the mesh during wet seiving. Fifty times this quantity is the total weight of mud in the cylinder. Since this may be directly linked to the known proportions of sand and mud in the entire sample this additional step enables the cumulative percentages of the total sample to be computed for given increments.

Computation of size percentage

In calculating the grain size distribution the percentage by weight of material finer than the nominal diameter for time t, is expressed as

$$F_t = \frac{M_t}{V_t} \Big/ \frac{M_o}{V_o} \tag{5.1}$$

where M_o is the original weight of the entire sample plus dispersant V_o the original volume (i.e. M_o/V_o is the original concentration), and M_t is the weight of sediment and dispersant in volume V_t taken at time t (i.e. M_t/V_t is the aliquot concentration). Allowance must be made for the presence of dispersant both in the sub-sample and in the original suspension. However, in the simplest mode of calculation the dispersant may be eliminated by determining the difference in the weight of material retained during two successive withdrawals. In each case the same proportion of dispersant D_a is dried with the sediment withdrawn in the same volume of suspension, so that the weight of sediment retained is given by $(m_{t1} - D_a) - (M_{t2} - D_a)$ or $M_{t1} - M_{t2}$.

In this way the proportion of coarse silt, medium silt, fine silt, etc., may be determined by measuring at the appropriate time intervals and the full size distribution established by continued repeated measurements until the smallest significant size in the sample is reached.

The proportion of sand (A) and mud (B) present in the wet sieved sample is determined. The weight of sediment remaining in suspension (C) at any given time is 50 times the weight of sample obtained at that time less dispersant. So it may be seen that the cumulative percentage of sediment coarser than any particular size is

$$\frac{100(A+B-C)}{A+B}\%.$$

Atterberg cylinder

The settling tube combined with a withdrawal system used in the Atterberg cylinder modified by Prokopovich (1958), entails the positioning of a siphon near the base of the tube. The suspension, prepared as for pipette analysis, is stirred, placed in the tube, and allowed to settle. After the lapse of a pre-determined fall time for a given equivalent diameter to have settled past the level of the siphon inlet, the entire mass of fluid above the siphon is drawn off, filtered, dried, and weighed.

By use of the Atterberg cylinder it is possible to separate well-defined grain fractions from a sample, by removing first the finest grades and later extracting progressively coarser fractions. Several repeated settlings of the residual sediment may be required before a total separation of a specific size is achieved. Details of the method are given by Müller (1967). The individual grain classes separated by this technique may be used in further investigation of the mineralogical or chemical characteristics of the fine sediments such as clay or fine silt. Although it may be used to provide a grain-size analysis the time required is up to one month, and the technique is best confined to the preparation of samples for other forms of examination.

Hydrometer analysis

Like the pipette method, hydrometer analysis depends upon the rate of settling of particles, but uses the change in density of the suspension as a measure of its content of solids (Müller, 1967; B.S. 1377, 1957). This technique has the advantage that all fine particles ultimately settle and that it involves no withdrawal of fluid or drying of suspensate. Variations of fluid density are reflected in the depth at which the hydrometer bulb floats below the fluid surface. Repeated introduction of the hydrometer generally leads to disturbance of the settling suspension, more so than in pipette analysis. In consequence many workers restir the suspension and permit it to restart sedimentation after each measurement. This greatly lengthens the duration of analysis. Other workers leave the hydrometer in the suspension throughout the analysis, thereby eliminating periodic disturbance, but particles may adhere to the instrument and so induce inaccuracies.

Objections to this method of size analysis are based on the facts that the hydrometer depth must be read from the top of the meniscus and a correction applied, that the method is strongly temperature-dependent, and that the hydrometer is placed in a fluid which exhibits a density gradient, but purports to indicate one single density, although the instrument itself extends through at least part of the density variation. In order to reach sufficient accuracy in reading the specific gravity, concentrations of over 40 g/l are required (Allen, 1968). Such concentrations virtually ensure that hindered settling will occur. The method is not sensitive to abrupt grain size changes and is not well suited to polymodal grain populations such as frequently occur in estuaries.

In its simplest form (Müller, 1967, *p*87) the method of calculation may be expressed as:

$$C_t = \frac{\rho_s}{\rho_s - \rho_1}(\rho_t - \rho_1), \tag{5.2}$$

where ρ_s is the density of the particles and ρ_1 is the density of the liquid used for sedimentation.

The full method of calculation including useful nomographs enabling corrections to be made for temperature, relationship of particle diameter to fall velocity, and calibration of the hydrometer is given in B.S. 1377 (1957). A valuable aid to the calculation is given by Buchan, Jones & Simpkin (1970) in the form of a computer program into which the hydrometer readings are inserted, and the size distribution is printed out.

Sedimentation tube

Both pipette and hydrometer involve some disturbance of the settling suspension and potentially introduce errors of handling during introduction and retraction of apparatus. In order to circumvent these problems, the use of sedimentation tube balance, in which particles are allowed to settle uninterrupted, has been developed. Sediment samples are introduced at the top of the tube and the particles settle to the other end at rates appropriate to their fall characteristics – dominated by size. Visual inspection of the sediment as it accumulates in the narrow, constricted basal exit tube (as in the Emery tube (Emery, 1938)) enables settling times to be established. In a second method the equipment automatically records the changes in pressure difference detected in the fluid between the top and the base of the tube, and that resulting from the weight of the sediment in the column above the lower transducer, and compares these with time (Schlee, 1966). A third method uses a balance pan situated within and near the base of the sedimentation tube to catch particles, and changes of weight retained with time are recorded (e.g. Sengupta & Veenstra, 1968). Settling techniques have also been used for coarse sediment (Channon, 1971).

The sedimentation tube is becoming increasingly used for the rapid determination of sand population sizes. Like pipette and hydrometer analysis it is based on Stokes' Law, which states that the drag force exerted on a sphere falling at low velocity is directly proportional to the product of the fluid viscosity, particle diameter, and the fall velocity. This is often restated as

$$V = \frac{2}{9} \cdot g \cdot \frac{\rho_s - \rho_1}{\mu} \cdot r^2, \tag{5.3}$$

which shows that the terminal fall velocity, V, is directly proportional to the density difference between the sphere and the liquid and to the square of the particle radius, and indirectly proportional to the liquid viscosity, μ.

Natural mineral assemblages rarely comprise spheres, but contain mixtures of many shapes of particle. Grains of the same composition and apparent radius settle at different speeds if they are of different shapes or have differing surface textures. Both of these properties influence the particle drag. Values for the spheres with smooth surfaces may be found in tables of constants, but corrections are required for irregularly shaped particles for similar projected diameters (Coulson & Richardson, 1966). The interpretation of fall velocities in terms of particle size, especially when many grains are released simultaneously leads to considerable problems.

It can be shown that within a cloud of settling particles in which the concentration exceeds one per cent, hindered settling occurs, and the sedimentation velocity is reduced by about 5% of the single particle fall velocity (Richardson & Zaki, 1954). Kuenen (1968) considered that for a 1-g sample settling convection does not become serious until the particles are less than 50 μ in diameter. Hulsey (1961) has shown that the fall velocities may be increased by 20% for samples of between $\frac{1}{4}$ g and 1 g released in sedimentation columns, but between 2 g and 5 g the changes were considerably less. Clearly optimum sample volumes which would avoid mutual interference between settling grains may be defined for each instrument.

Thus the quantity of materials used determines the fall characteristics, e.g. whether the grains fall as bunches or in a uniform spread. Since small quantities of sediment are required for most analyses, problems could arise in the selection of sub-samples as being representative of the whole sample. However, the analytical procedure is relatively quick, and so this problem is partly overcome by running several sub-samples from each sediment to be analysed.

Although precision, represented by close similarity of results achieved by repeated analysis of the same samples may be high, the interpretation of the results is less certain. It is possible to calibrate the system against curves obtained for single spheres (Zeigler & Gill, 1959), or clusters of spheres of known single diameters using glass spheres or ballotini (Schlee, 1966), but the curves obtained are relevant only to similar concentrations of spherical grains of identical density. Natural sediments comprise many irregularly shaped particles of differing densities and surface textures, so that full calibration for all possible combinations is quite out of the question. Comparison of sieve and sedimentation column analysis has been carried out by Sengupta & Veenstra (1968) and by Sanford & Swift (1971) who concluded that the two methods yielded very similar overall grain-size data but that the tube analyses overestimated the particle size of fine samples and underestimated the diameters of the coarse samples.

In addition to the problems outlined above the introduction of the sample at the head of the tube may lead to difficulties in ensuring the simultaneous release of all particles. This is normally achieved by the use of some wetting agent on the platen surface. The column of standing water develops internal instabilities through the natural establishment of convection currents. In the absence of temperature controlled surroundings these currents disrupt the static water in the 2 m high sedimentation column.

Coulter counter

Where available the use of a Coulter Counter system provides a very detailed analysis of the grain sizes present in a sediment sample. The method is described in Chapter 6. The method is subject to errors relating principally to the response time of the electrical impulses and their separation, but nevertheless it probably provides the best continuous size-determination method currently available. Analysis is simple and rapid, and once fully calibrated this expensive equipment is capable of producing many detailed size distributions in a short period of time (McCave & Jarvis, 1973). Despite the evident high value of this form of analysis, which is widely used in industry, the equipment is rarely available in sedimentology laboratories, and most sedimentologists are forced to rely on the sieving and pipette methods for the analysis of estuarine sediments.

Analysis errors

As has been shown there are many techniques for the determination of particle size in sediments. It is important that the reproducibility achieved with any particular technique should be assessed by means of repeated analyses of the same material. Once the errors inherent to the apparatus or the method have been determined in this way the investigator may proceed with the study knowing the limits which may be placed upon the reliability of the analytical results. Before these tests are carried out it is essential that the full analytical procedure should be performed many times to ensure that the operator is fully conversant with the methods and automatically adopts a standardised procedure.

As indicated, no single method is appropriate for all grain-size determinations. At the limit of their optimum ranges the results obtained by the various techniques decrease in accuracy. In consequence it is quite common to find apparent deficiencies or excesses of particles in such sizes where more than one technique has been used to establish the grain-size distribution. Some such deficiencies may really exist in nature, e.g. particles 1–2 mm in diameter may not become permanently deposited until they have undergone size reduction through inter-particulate collisions. In addition to those naturally occurring, deficiencies or excesses may appear as a result of flaws in the methods of analysis. For example determination of particles in the range 40–60 μm is achieved by pipette methods. The time interval between thoroughly stirring the sample and withdrawal of the first sub-sample is small and residual currents induced by the stirring may continue to buoy up small particles for much of the period. In consequence there appears to be less material in this size range than is actually present.

In a consideration of nearly 7000 size analyses Griffiths (1967) demonstrated deficiencies of fine gravels and silts in general. He concluded that statements of overall lack of particles of given sizes may be more a measure of sampling bias introduced because investigators were interested principally in features other than the grain size, such as petrography of clays, shape and fabric of gravels, or beach sand variability.

Grain-size distribution

The style in which the raw data are recorded depends on the analytical technique used. In sieving, the weight of sediment retained on each mesh is logged (Table 5.2, col. 5). This is then converted to a percentage of the total weight of sediment analysed (Table 5.2, col. 7). The conversion of settling velocities to grain diameter equivalents is more complex, but a similar percentage frequency table may be produced. These tables provide the basic information which the sedimentologist manipulates in order to obtain various characteristics of the sediments (Fig. 5.4*a*). The simplest mode of presenting a statement on the relative abundance of various grain-size groupings is by means of a ternary diagram in which the total sediment at any site is divided into three fractions: gravel, sand and fines on a percentage basis. The sediment may then be represented by one point on a triangular graph for which each apex serves as a sediment entirely of gravel, or entirely of sand, etc. The scales of subdivision of the sides of the equilateral triangle may be arithmetic, or geometric. A series of fields of sediment types may be erected (Folk, 1974) and verbal descriptions of the sediments and their geographical distribution within the estuary thereby rendered possible.

Histograms

The simplest method of presenting the data in graphical form is by means of histograms. In these the independent variable is grain size, which is therefore related to the horizontal axis, and the dependent variable is the weight percentage, which appears on the vertical axis. The percentage of sediment retained on a specific sieve may be plotted as a bar at that size, or as a column rising to that percentage height and extending from the sieve mesh to the next coarser sieve mesh used. The number of columns produced depends on the number of grades chosen or calculated during grain size determination, and also on the grain-size spread of the sample (Fig. 5.4*b*).

Construction of graphs is simplified if the size scale consists of equal increments. As will be seen from Table 5.2 the sieve meshes commonly used do not follow an arithmetic progression, but a geometric one, for small

a

Wentworth grades	Grade intervals	Frequency weight (%)	Cumulative frequency weight (%)
Granule	−2 to −1 ϕ 4 to 2 mm	1	1
V. coarse sand	−1 to 0 ϕ 2 to 1 mm	7	8
Coarse sand	0 to 1 ϕ 1 to 1/2	14	22
Medium sand	1 to 2 ϕ 1/2 to 1/4	23	45
Fine sand	2 to 3 ϕ 1/4 1/8	28	73
V. fine sand	3 to 4 ϕ 1/8 1/16	18	91
Silt	4 to 5 ϕ 1/16 1/32	7	98
	5 to 6 ϕ 1/32 1/64	2	100

b

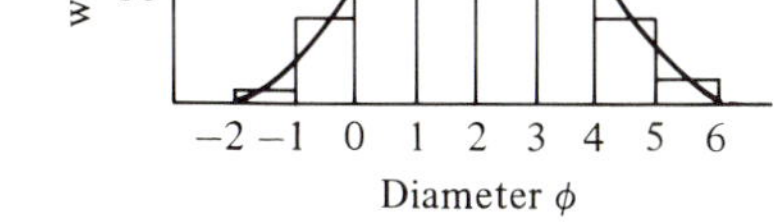

c

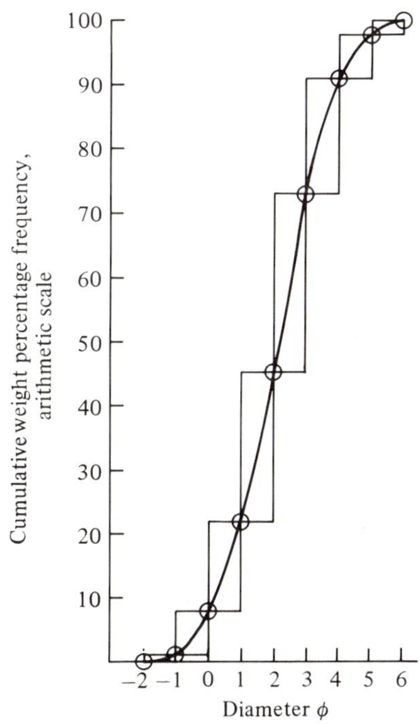

Fig. 5.4. *a*. Table to show the analysis of a sediment according to the percentage of weight falling into each grade size, and the cumulative frequency data. *b*. Plot of frequency by weight percentage against grain diameter using data from the table. *c*. Plot of cumulative percentage frequency by weight on an arithmetic scale against grain diameter for the same data.

variations of dimension of a large grain are much less important to its behaviour than are identical variations to a fine particle. In consequence it is normal to analyse sediments with respect to a scale which follows a geometric progression. The scale may be subdivided into equal length segments which correspond to the size gradations suggested by Wentworth (1922), who recognised the importance of the deviation from the arithmetic distribution. The divisions are referred to as forming a 'phi scale' following Krumbein (1934) who realised that Wentworth's scale could be written logarithmically using the conversion

$$\varphi = -\log_2(\text{grain diameter in millimetres}).$$

In one sense the frequency curve is a histogram whose grade intervals have become so small that the column widths tend to zero, and the columns can be represented by an infinite number of adjacent vertical lines. The frequency curve is the smooth curve joining the top of these lines, and shows the variation of weight percentage as a continuous function of grain size. Such curves may be achieved using Coulter Counter techniques, but the approximation of the frequency curve by other methods depends on the number, coarseness, and constancy of the grade intervals. Detailed precise frequency curves may be constructed, but such curves are in practice rarely used, and most workers form them by superimposing a smooth curve over the histogram columns (Fig. 5.4.*b*).

Cumulative frequency curve

The raw data for the construction of the histograms and frequency curve may be used also in the construction of the *cumulative frequency curve* (Table 5.2, col. 9). In this case the vertical axis is laid off in divisions from 0 to 100% and the horizontal axis retains its size scaling. The percentage weight retained on the coarsest sieve is plotted at the relevant diameter, the sum of the sediment weight percentage in the coarsest two sieves is plotted above the next finer sieve mesh size, the sum of the residue on the coarsest three sieves above the third value, and so on, until the entire sediment distribution has been accounted for at the cumulative value of 100%. The entire procedure is akin to stacking the histogram columns one above the other in a stepwise progression (Fig. 5.4*c*).

The ordinate scale used for cumulative frequency distribution curves may be varied from the arithmetic one presented in Fig. 5.4*b*, *c*. Most sediment grains in naturally occurring populations cluster about a given size with progressively fewer grains finer and coarser. The distribution is a form of normal probability distribution, a curve which has a bell shape. The probability scale is introduced to produce a straight line plot for samples which show a true symmetrical probability distribution of grain size (Fig. 5.5). This scale compresses the central section and extends the very fine and very coarse 'tails' of the distribution. Plotting cumulative frequency distributions on such a scale straightens the S-shaped curve of the arithmetic ordinate scale, reveals any departures from the normal probability law, and makes it easier to interpolate and extrapolate percentile values which may be required for statistical characterisation of the sediments.

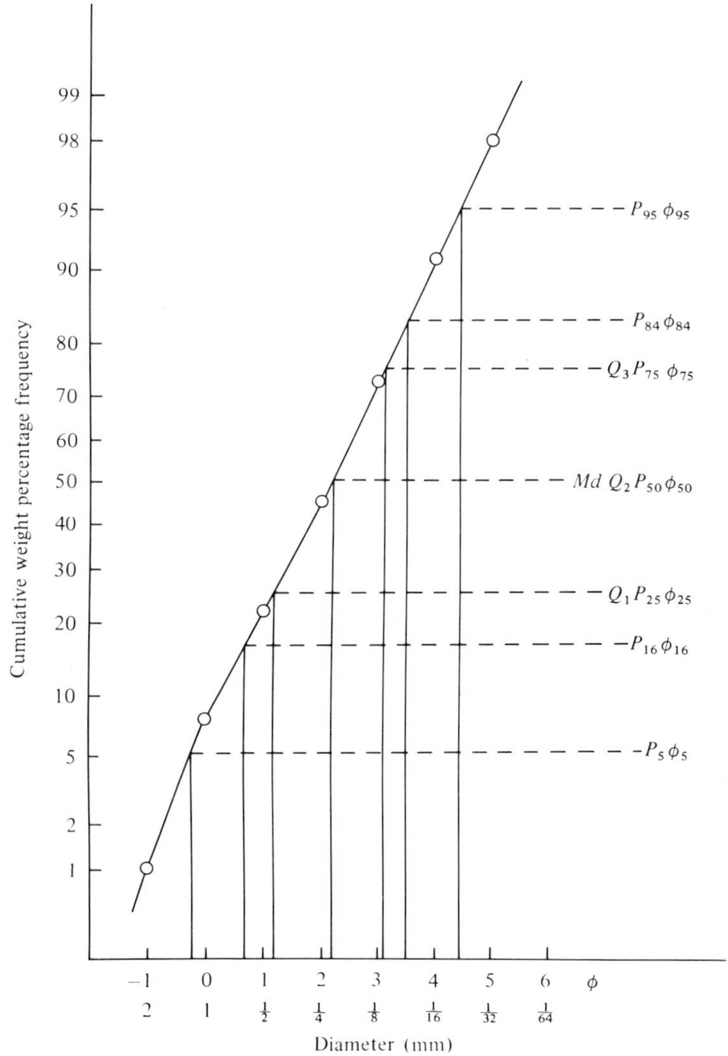

Fig. 5.5. Plot of grain-size distribution as cumulative weight percentage frequency against grain diameter using a probability ordinate scale. The main percentile values used in graphic statistics are indicated with their normal abbreviations.

Graphical presentation

The strength of the histogram lies in its illustrative impact. Trends in the frequencies of individual size grades may be seen at a glance, and it is especially effective for the identification of primary and secondary modes (most common grain sizes) and their intervening saddles (Fig. 5.6B, C). By comparing histograms the analyst can vividly recognise sequential changes

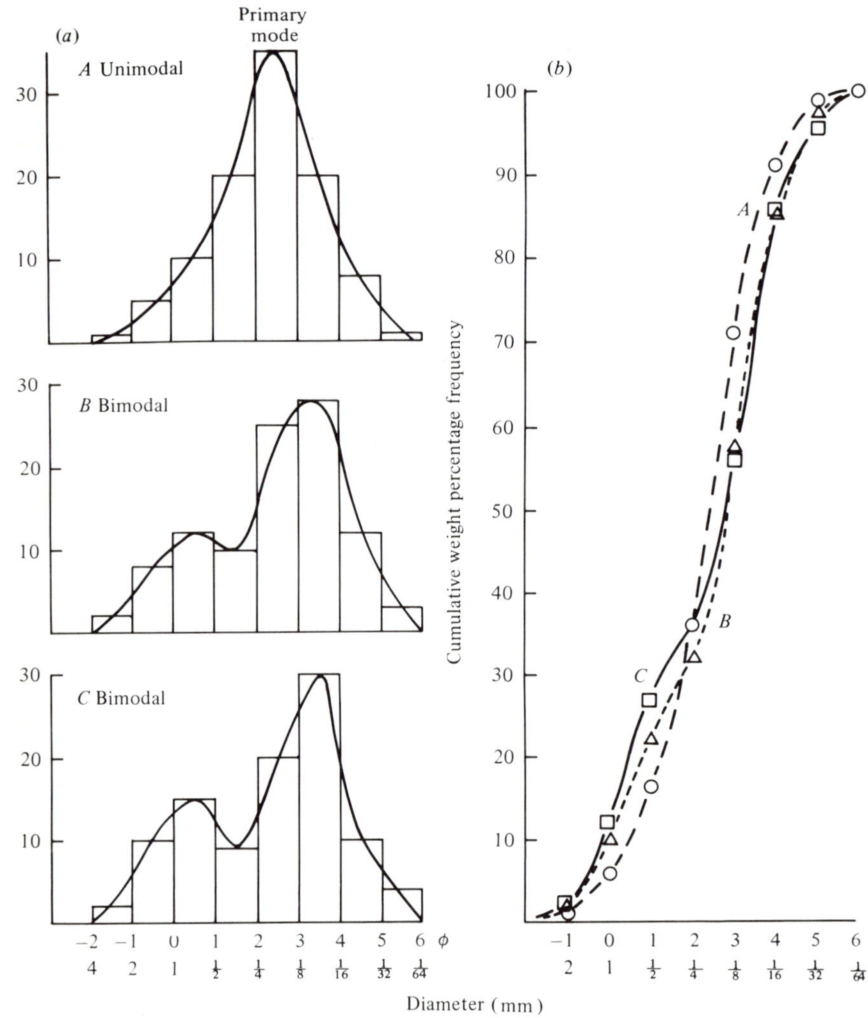

Fig. 5.6. *a*. Plots of frequency as weight percentage against grain diameter. *A*, unimodal; *B*, weakly bi-modal; *C*, strongly bi-modal sediment. *b*. Plots of cumulative weight percentage frequency against grain diameter for the same sediments as in *a* using an arithmetic ordinate scale.

which may not be adequately described by statistical measures, but which are readily characterised by words. For example, assume that Fig. 5.6*a* represents analyses from a related series of sediments. Sample *C* largely consists of a mixture of coarse and fine sand with lesser proportions of granules, medium sand and silt. The overall impression is that the sediment is poorly sorted, skewed towards the 'fines', and is strongly bimodal, with the primary mode displaying a relatively sharp peak. Sample *B* differs in degree: the bimodality is weaker, the peaks are less sharply defined, and the dominance of the very fine sand grade ($\frac{1}{8}-\frac{1}{16}$ mm) is being matched in frequency by the fine sand grade ($\frac{1}{4}-\frac{1}{8}$ mm). In sample *A* the coarse secondary mode has disappeared, the primary mode is centred on the fine sand grade, the overall distribution is symmetrical, the sediment is better sorted, and the primary mode forms a sharp peak. This information has been gleaned without recourse to statistical calculations.

The strength of the frequency curve depends on its method of construction. If the 'short cut' technique is employed then its value is similar to that of the histogram, i.e. illustrative, with the single advantage that it is a smooth curve and gives the impression that the data are continuous. Its major advantage lies in its ability to define accurately modal values which cannot be realised satisfactorily in any other way.

To achieve this the curve must be constructed from the cumulative frequency curve so that its final form is truly continuous and independent of grade scales. The graphical transformation to yield the frequency curve is based on the measurement of tangents at various points along the cumulative curve, especially at points of inflection. The procedure is not difficult, but it is lengthy and cannot be described in this brief chapter. Instead we refer the reader to the descriptions by Krumbein & Pettijohn (1938, pp. 191–5) and Folk (1974, pp. 42–3).

Of the three graphical styles discussed the cumulative curve is the least pictorial. In Fig. 5.6*b*, three curves have been constructed using the same data as those for the histograms of Fig. 5.6*a*; yet only the most experienced eye could read the same information from them. Certainly it is possible to compare cumulative curves in terms of their slopes (a function of sorting), their inflections (modes), and their degree of asymmetry, but their overwhelming attribute is not illustrative; they are ideal for the extraction of percentile values which form the data for quantitative statistical characterisation and analysis.

Statistical characterisation

A first encounter with sedimentary statistics can be a daunting experience. Bewilderment arises from the plethora of measures available, many with deceptively similar names, and the fact that whereas some are directly analogous to those of conventional statistics others are not.

One way in which sedimentary statistics differ from conventional statistics of normal distributions is that frequency is normally expressed as a weight percentage rather than as a number. Two scales of size are available, metric or phi-based. The phi scaling forces the distribution of sediment size into an approximately normal distribution. There are two main ways of obtaining statistical information from sedimentary data. The first is to extract selected percentile values from cumulative curves and feed them into *graphic* statistical formulae: this is the most popular method. The second is to use the *method of moments* whose measures are influenced by every grain in the population.

The common ground is that each method attempts to characterise sedimentary grain size data using the most standard, conventional statistical analogues. These include:

(a) measures of central tendency, or averages (median, mean, mode), designed to provide values about which all other values cluster;

(b) measures of the degree of scatter (dispersion, deviation, sorting, spread), to provide values which indicate the spread of data about the average;

(c) measures of the degree of symmetry (skewness) to provide a measure of the tendency of the data to spread preferentially to one side of the average; and

(d) measures of the degree of peakedness (kurtosis) of the distributions.

In combination these four sets of measures provide a comprehensive characterisation for any distribution.

Graphic methods. The figures used in the calculations are read as the values corresponding to given percentage values on the cumulative frequency curve. The Nth percentile of size is the size of particle for which $N\%$ of the sample is finer and $100-N\%$ is coarser. For example in Fig. 5.4 the 5th percentile has a diameter value of 1.1487 mm or -0.2φ, so that 5% of the sample is coarser and 95% is finer than this size. Similarly the diameter value of the 50th percentile is 0.2176 mm or $+2.2\varphi$; the 95th percentile is 0.0474 mm or $= +4.4\varphi$.

Various examples of *graphic* measures using percentile inputs are listed in Table 5.4. Folk (1966) has reviewed the various measures. In general there are two fundamental differences between the metric and phi-based formulae: first the length of the distribution curve embraced by a set of measures, and secondly, the interpretation of phi versus metric statistical values.

With the exception of Kq_a (Kurtosis) which is not in common use today, all of the *metric* measures employ only the 25th, 50th and 75th percentiles. Their coverage is confined to the central 50% of the distribution curve, the

Table 5.4. *Some common statistical measures using the metric and phi unit systems*

System	Median	Mean	Dispersion deviation sorting	Skewness	Kurtosis
Metric	$Md = Q_2 = P_{50}$	$M = \frac{Q_3+Q_1}{2}$	$QD_a = \frac{Q_3-Q_1}{2}$	$Sk_a = \frac{Q_3+Q_1-2Md}{2}$	$Kq_a = \frac{Q_3-Q_1}{2(P_{90}-P_{10})}$
			$So = \left(\frac{Q_3}{Q_1}\right)^{\frac{1}{2}}$	$Sk = \frac{Q_3 Q_1}{Md^2}$	
Phi	$Md = \varphi_{50}$	$M_\varphi = \frac{\varphi_{16}+\varphi_{84}}{2}$	$\sigma_\varphi = \frac{\varphi_{84}-\varphi_{16}}{2}$	$\alpha_\varphi = \frac{M_\varphi - Md_\varphi}{\sigma_\varphi}$	$\beta_\varphi = \frac{\frac{1}{2}(\varphi_{95}-\varphi_5)-\sigma_\varphi}{\sigma_\varphi}$
		$M_z = \frac{\varphi_{16}+\varphi_{50}+\varphi_{84}}{3}$	$\sigma_I = \frac{\varphi_{84}-\varphi_{16}}{4} + \frac{\varphi_{95}-\varphi_5}{6.6}$	$Sk_I = \frac{\varphi_{16}+\varphi_{84}-2\varphi_{50}}{2(\varphi_{84}-\varphi_{16})} + \frac{\varphi_5+\varphi_{95}-2\varphi_{50}}{2(\varphi_{95}-\varphi_5)}$	$K_G = \frac{\varphi_{95}-\varphi_5}{2.44(\varphi_{75}-\varphi_{25})}$

Md, Median; *M*, mean; QD_a, arithmetic quartile deviation; *So*, Trask sorting coefficient; Sk_a, arithmetic quartile skewness; *Sk*, geometric quartile skewness; Kq_a, arithmetic quartile kurtosis (Trask, 1930, 1932; Krumbein & Pettijohn, 1938; Krumbein, 1939); M_φ, phi mean diameter; σ_φ, phi deviation measure; α_φ, phi skewness measure; β_φ, phi kurtosis measure (Inman, 1952); M_z, mean size; σ_I, inclusive graphic standard deviation; Sk_I, inclusive graphic skewness; K_G, inclusive graphic kurtosis (Folk & Ward, 1957).

area between the third and first quartiles (Q_3 and Q_1), and ignores the 'tails'. The phi-based formulae extend the coverage, first with a bias on the central 68% of the distribution (Inman, 1952); and then to 90% of the distribution (Folk & Ward, 1957).

The choice of coverage is still open to debate. One school of thought states that the 'tails' of a distribution are the most sensitive indicators of a sediment's depositional environment, and that the statistical characterisation which reflects the conditions to which they have been subjected must embrace them. Another school suggests that the 'tails' are unreliable because they reflect different depositional mechanisms. The tails also reflect the errors in the analysis techniques.

For example, Visher (1969) has demonstrated that the coarsest 20% of a distribution may have been the traction load, while the finest 20% settled out of suspension. If it can be assumed that the central population (saltation load) represents the dominant mechanism, then the 50–60% of a grain-size distribution curve is the most likely segment to reflect the integral nature of the *principal* depositional mechanism in an environment.

However some estuarine sediments are bimodal, composed entirely of a coarse deposit with fine material derived from suspension within the matrix and with no evidence of a bedload component. The metric measures may provide a descriptive framework for such material, but on their own would not reveal the disparate nature of the sediment. Analysis using the central segment would be equally inappropriate.

Irrespective of the finer points of coverage, the problem is partly solved by the nature of estuarine sediment samples themselves. Coarse sands are invariably intermixed with granules, pebbles, shelly detritus and wood fragments. Medium sands may contain mud pellets, peat fragments and leaves, and fine grained sediments may contain much organic matter, faecal pellets, shell fragments and other 'impure' components. Of course, many of the larger components can be removed by hand-picking or chemical dissolution, and finer aggregates separated by chemical dispersants. There are limits however, to this form of tampering before the *absolute* estuarine character of the sediments is completely destroyed in deference to the sizing technique (see Buller & McManus, 1975). The resulting compromise, which can only be based on experience, ensures that the essential estuarine elements are preserved, but many open-ended size frequency distribution curves will be produced. In practice, the percentile range between 16% and 84% is the largest possible if the majority of samples are to be characterised by the same measures. Of those shown in Table 5.4, the metric forms, M, QD_a, So and Sk could be used; and the phi measure M_φ, σ_φ and α_φ would also be suitable.

The merits and interpretation involved in using metric or phi-based measures are also subject to debate, and only a simple account is given below.

First, it must never be forgotten that when a phi system is used a frequency curve represents the unique relationship between frequency as weight percentage, versus $-\log_2$ of the grain diameter in millimetres. When the metric system is used a frequency curve represents the unique relationship between frequency as weight percentage and the grain diameter in millimetres. The style in which the unique frequency curve is presented (for example, as a cumulative frequency curve plotted on arithmetic-probability graph paper with a phi abscissa) has absolutely no influence on the values of statistical parameters characterising it. It is simply a graphical convenience for extracting percentiles. For these reasons statistical values calculated from the phi- or metric-based measures which employ various *combinations* of percentiles are not directly interchangeable: Recall that the sum of two logs is equal to the log of their product, and the difference is equal to their quotient. There is, however, one exception to this rule. The median is equivalent to the 50th percentile, and an *individual* percentile value is directly interchangeable from phi to millimetre units.

The simplest attitude is to accept that the two unitary systems of measures are entirely different, and to assess them on their separate merits.

The values of the metric quartile measures M, QD_a and Sk_a all depend on the size of grains involved and the results are dimensional (in millimetres). The equation for M and QD_a are self explanatory (Fig. 5.6; Table 5.4) while Sk_a represents the departure of the mean (M) from the median (Md). The geometric measures So and Sk are size independent and dimensionless because they are based on ratios rather than sums or differences. The arithmetic quartile kurtosis Kq_a compares the spread of the central 50% of the curve to the spread of 90% of the curve. It, too, is independent of size and is dimensionless because it represents the ratio between two spreads. Of the entire suite, only Md, QD_a and Sk_a are in popular usage today.

The phi equivalents of the metric measures M and QD_a are conceptually and mathematically similar in formulation although their percentile coverage extends well beyond the 25th and 75th limits (Table 5.4). The difference between M and M_φ or M_z is simply that the latter involve logarithmic sums, the results of which are expressed in phi. The difference between QD_a and σ_φ is, however, of fundamental importance. The results are also in phi, but in this case they represent the spread of a curve which can be expressed as a certain number of Wentworth grades. This concept is a major departure from the metric system because the spread of Wentworth grades, or the number of phi divisions, is independent of the median and mean. For example, a result of 4φ might represent a spread from granules to medium sand; or equally, a spread from the fine sand to clay. Without the median or mean value there is no way of telling where the spread is centred in the size scale.

The phi equivalents of the metric measure Sk_a differ in three ways: α_φ and Sk_1 are geometrically independent of sorting, the results are dimensionless because phi spreads occur in both the numerators and denominators, and phi spreads are median and mean independent. Similar comments apply to the phi kurtosis measures of β_φ and K_G.

Of the entire suite the Folk & Ward (1957) parameters M_z, σ_I and Sk_I are widely used, but for open ended distributions Inman's (1952) M_φ, σ_φ and α_φ are more practical.

The question of which system to use is still a matter of conjecture. A trite answer such as 'the one which provides the most meaningful results for your particular purpose' is woefully inadequate, but without invoking dogmatism it is the best that can be offered. The purists prefer phi measures because to them the phi system enshrines the logarithmic distribution of sedimentary grains; it provides a neat link between phi, the *unit* and the phi, the *spread*, which represents one Wentworth grade, and furthermore, the measures are largely independent one from the other. The 'realists' suggest that the average size, sorting and skewness of sediments are integrally related in nature, and that the statistical characterisation should reflect this integration. The efficiencies of some phi-based statistical measures in describing the distribution have been compared by McCammon (1962).

Method of moments. The method of moments is more closely related to 'orthodox' statistics. It uses the entire population of grains in the sample and may be applied only in cases in which the size distribution has been fully determined, with no grains of either larger or smaller diameter than those actually measured. In other words, it is applicable only for 'closed' distributions, and in this respect it contrasts with the graphic techniques which may be used for 'open-ended' distributions, where there are appreciable proportions less than the smallest sieve size. The statistical measures obtained are influenced by every grain in the sediment sample rather than by just a few selected percentiles of the distribution.

Since many estuarine sediments have a clay fraction and show open-ended distributions, the method of moments is not of universal application in estuarine sedimentology. However, in some sub-environments, such as sandy inter-tidal flats and entrance beaches, it may be suitably used. It must be stressed that the practice of extrapolating distribution curves at their extreme ends to produce an artificially closed distribution may introduce severe errors, for the more remote grains are from the centre of the distribution the more disproportionate their effect upon the statistical measures derived.

The method of moments, introduced into sedimentology by Krumbein (1934) has been applied increasingly in the past decade because computers have made the calculations easier to carry out. It provides measures of average grain size (first moment), variance (second moment), and the skewness and kurtosis of the distributions (derived from the third and fourth moments). The full calculation procedures are set out in texts such as Krumbein & Pettijohn (1938), Griffiths (1967) and Folk (1974), to which the reader is referred.

Presentation and interpretation of results

As indicated earlier in this chapter, the many uses to which sediment analyses may be put each require specific factors to be high-lighted. For most purposes general statements concerning the muddiness or sandiness of the sediment are precursors of further, detailed descriptions.

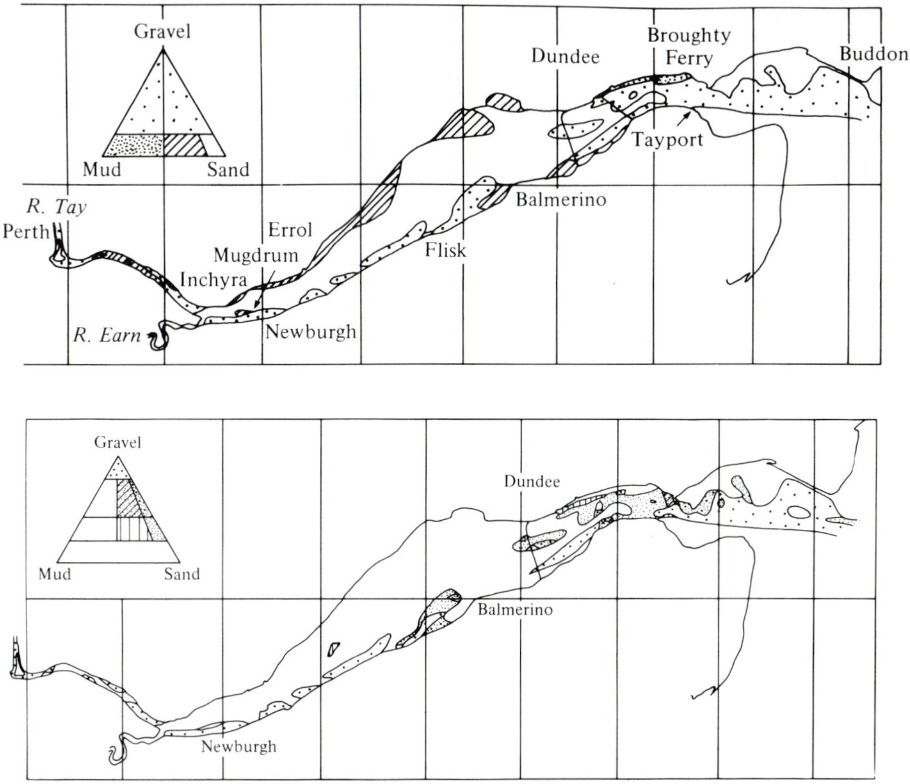

Fig. 5.7. *a*. General distribution of sediments in the Tay Estuary as indicated by a ternary diagram in terms of gravel, sand and mud. (After Williams & McManus, 1975.) *b*. Detailed subdivision of the coarse sediments in the Tay Estuary using the same ternary diagrams as in *a*.

The distribution of sediment types throughout the estuary according to the fields on the ternary diagram may be plotted. For example, in Fig. 5.7*a* the sediments of the Tay Estuary are related to such a ternary diagram. Coarse field subdivision has been made, but nevertheless the major part of the estuary bed is seen to be dominated by sands with little variation of size. Muddy sands dominate the tidal flats of the upper reaches and fluviatile areas, and various forms of sandy or muddy gravels characterise the main channel which hugs the south shore west of Balmerino. The seaward reaches are largely of gravelly sands, and the North Sea beaches are again sandy. Further subdivision (Fig. 5.7*b*) enables variations of the gravels to be detailed in a form suitable for assessment of the distribution of exploitable aggregate, by concentrating attention on this particular part of the field.

For many estuaries this form of sediment distribution presentation has been made. The results have been used in a sedimentological context by Nichols (1972), in connection with mineralogical studies in four United States estuaries (Folger, 1972) and by Palmer (1974) in Chesapeake Bay, with geochemical investigations in the Rio de la Plata (Biscaye, 1972; Urien, 1972) and the Genoa estuary in Australia (Reinson, 1975), with biogeochemical surveys in the Rappahannock Estuary (Nelson, 1972), in radioactive tracer studies (Martin, 1970), and to indicate the distribution of frictional resistance due to bed material during tidal wave modelling (Williams & McManus, 1975).

The use of the frequency curve is not widespread in analysing sediment distribution patterns. Nevertheless it can be seen from Fig. 5.8 that the relative importance of specific modal groups, and their decay as mixing of populations takes place, may be used to give an indication of sediment movement paths within the estuary. Three initial sediment mode groups are interpreted as entering estuary: the coarse lag material rarely moved by currents, and a fine suspension load (both from the land), and a medium grained, mainly saltation-moved load derived from the sea. Typical changes of population modes are seen in Fig. 5.8*a*, and the effect of these on cumulative frequency curves is shown in Fig. 5.8*b*. The coarse and fine modes characterising the river typify much of the upper estuary and channel, but sands of the central mode are widespread at the seaward end. As the marine sands reach the tidal flats they are mixed with existing bimodal populations near the channel, but as the coarse material is unable to move far from the channel the sediment loses its trimodal nature, assumes bimodality based on the medium and fine modes, and ultimately in the quietest areas becomes unimodal, when it is entirely of fine sediment.

Extraction of sediment characteristics based on graphic statistical measures may enable variations of such factors as median, sorting, or skewness to be

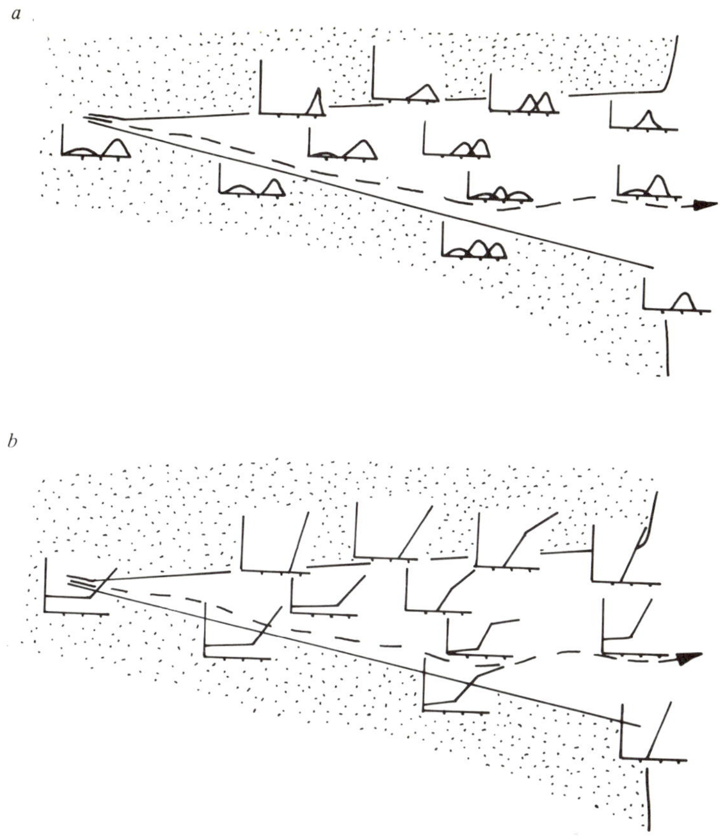

Fig. 5.8. Variation in sediment populations in an idealised estuary. *a*. Frequency plots of sediments from different localities related to three input population modes. The coarse and fine populations originate from the river; medium grained populations from the sea. On the plots coarse material is to the left, fine to the right. *b*. Cumulative frequency percentage by weight plots using probability scale for the sediments of localities in *a*. This illustrates some variations possible using three modes in various combinations. In practice there may be many more modes and more complex inter-relationships.

plotted on an areal basis. These maps yield very different items of information from those hitherto considered.

In Fig. 5.9*a* the median distribution for Tay Estuary sediments (Buller & McManus, 1975) conforms closely with topography. The coarser sediments are restricted to the major channels, and progressively finer materials dominate the the extensive north-bank tidal flats.

The sorting (Fig. 5.9*b*) σ_φ is essentially independent of the median. However, in the Tay Estuary the best sorted sediment corresponds to areas

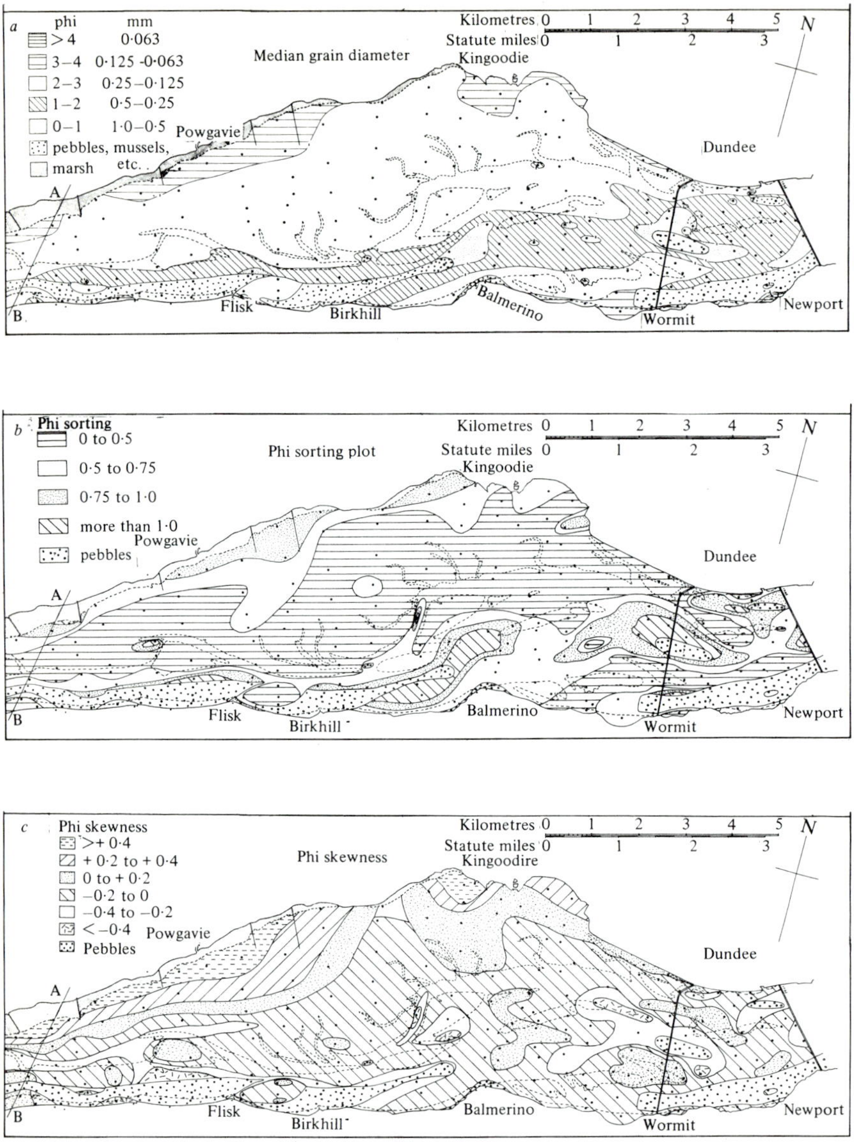

Fig. 5.9. *a*. Distribution of median grain diameter in part of the Upper Tay Estuary. *b*. Sorting characteristics of sediments in part of the Upper Tay Estuary. *c*. Skewness variations of sediments in part of the Upper Tay Estuary. After Buller & McManus, 1975.

of fine sand, with a decrease of sorting into the finer and coarser sediments which border the broad central swathe. Thus, sorting of channel sediments is seen to be as poor as that on the uppermost tidal flats.

Skewness, α_φ, is a dimensionless measure. Positive values indicate distributions which are dominated by the finer grades, and negative values reflect dominance of coarse grades. In the Tay Estuary (Fig. 5.9*c*) the upper tidal flat sediments are strongly positively skewed, the distributions are symmetrical lower on the flats, and become weakly negatively skewed on approach to the channel. Many sedimentologists have observed that negative skewness is often associated with sediments deposited in environments dominated by wave activity or strong current action.

Each of the statistical measures may be plotted against position in the estuary or, since many measures are interlinked, it is also useful to plot them against each other to show areal distribution of distinct fields of sediment textures. (It is important to realise that the greater the range of percentiles demanded for the derivation of statistical measures, the more restricted becomes the number of known points because so many of the estuarine sediment distributions are open-ended.) One example of this form of treatment relates the three quartiles of the cumulative frequency curve (Doeglas, 1971). Other examples of this method of presenting data are given in a review by Buller & McManus (1975).

Although the examples quoted here are derived principally from the Tay Estuary, with which the authors are most familiar, published maps showing the distribution of similar measures are available for many other estuaries. Comprehensive sediment maps by Allen (1971) for the Gironde and by Ryan & Goodell (1972) for Mobile Bay are typical examples.

The methods so far considered enable comparison of individual measures, or combinations of measures, but modern advances in sedimentology have led to the introduction of non-parametric statistics to compare entire grain size distributions. Principal Component Analysis examines the covariation of groups of properties which may include various organisms, heavy mineral populations, grain size and statistical measures. R-mode factor analysis (Klovan, 1966, 1975; Harbaugh & Merriam, 1968; Allen *et al.*, 1970) enables variation of entire grain size populations to be examined and indicates the interrelationships of various parts of the populations in space. The results of such grouping techniques require the investigator to suggest the dynamic, chemical or biological factors determining the distributions that appear. An example of such a study in an estuarine geochemical survey is given by Reinson (1975).

Following an entirely different line of approach Passega (1957) suggested the sediments could be characterised by the ratio of the coarsest one percentile, *C*, to the median diameter of the population, *M*. He reasoned that the physical processes controlling deposition would give an average grain size, and any extreme variation of current would be indicated by the coarsest grains deposited. As pointed out above the use of tails in any distribution, which are subject to the greatest errors of determination, may demand extrapolations, or be quite inapplicable to coarser sediments. Nevertheless Passega recognised that most sediments fell within a well-defined framework or a series of envelopes, characterising deposits from tractive currents or sediments from beaches. He showed that segments of his analytical graph could be related to deposits from uniform suspension, from graded suspension, or mainly from rolling of the particles on the bed. Thus he attempted to construct a texture-process discriminant relationship. In later papers Passega (1964, 1972) extended and refined the techniques, and demonstrated a method of distinguishing between marine and fluviatile deposits (Passega, 1974).

However, most researchers consider that the sediment populations are composite in nature and are formed of more than one subpopulation. Doeglas (1946) attempted to link the individual subpopulations to different transport conditions.

In detailed studies of the nature of sands and pebbles in a range of environments Moss (1962) demonstrated that there was systematic variation of particle shape with size. Three distinct size–shape populations were recognised and shown to be environmentally or process-sensitive. He suggested that all three populations could be distinguished in the cumulative curves of many naturally occurring grain assemblages. They comprised a dominant saltation load with subordinate traction and suspension loads. In a second contribution Moss (1963) suggested the forms of particle motion which contributed to the development of the subpopulations, linking his analyses with dynamic controls for waterlain deposits.

Studies of subpopulations were extended by Visher (1969) who suggested that multiple straight line segments could be distinguished in natural grain size distributions when plotted on probability paper. He considered that each straight line segment represented a subpopulation of normal distribution, and that each of these could be related to a specific mode of transport and deposition. In an analysis of the sediments of the Altahama estuary, Visher & Howard (1974) suggested the forms of subpopulation and correlated these with local dynamic processes. The process–frequency distribution link was taken further by Sagoe & Visher (1977) who attempted a deeper analysis of the controls by fluid dynamics.

In this chapter the authors have attempted to summarise the methods of sampling sediments, processing them and extracting various facts about their textures. It remains to reiterate that there is no universally useful sediment characterisation technique. Each reveals some distinct property or group properties whose variation is of greatest significance to the problem being studied. From that point onwards the interpretation of variations in other factors may be sensibly shown to be related or not related to sediment texture as the case may be.

REFERENCES

Allen, G. P. (1971). Relationship between grain size parameter distribution and current patterns in the Gironde Estuary (France). *J. Sed. Petrol.* **41**, 74–88.

Allen, G. P., Castaing, P., Feral, A., Klingebiel, A. & Vigneaux, M. (1970). Contribution à l'étude des facies de comblement et interprétation paléogéographique de l'évolution des milieux sédimentaires et actuels de l'estuaire de la Gironde. *Bull. Inst. Géol. Bassin Aquitaine*, No. 8, 99–155.

Allen, T. (1968). *Particle size measurement.* Chapman & Hall Ltd, London.

Biscaye, P. E. (1972). Strontium Isotope Composition and Sediment Transport in the Rio de la Plata Estuary. *Geol. Soc. Amer.*, Mem. 133, 349–58.

British Standard (1975). *Methods of the test for Soils for Civil Engineering purposes.* BS 1377: 1975. British Standards Institute, London.

Buchan, J., Jones, A. & Simpkin, P. (1970). A computer program for the calculation of hydrometer size analysis. *Mar. Geol.* **9**, M23–M29.

Buller, A. T. & McManus, J. (1975). Sediments of the Tay Estuary. 1. Bottom sediments of the Upper and Upper Middle Reaches. *Proc. Roy. Soc. Edinburgh.* (B), **75**, 41–64.

Carver, R. E. (1971). *Procedures in sedimentary petrology.* Wiley Inter-Science, New York.

Channon, R. D. (1971). The Bristol fall column for coarse sediment grading. *J. Sed. Petrol.* **41**, 867–70.

Coulson, J. M. & Richardson, J. F. (1966). *Chemical Engineering* Vol. 2. *Unit operation.* Pergamon Press, Oxford.

Doeglas, D. J. (1946). Interpretation of the results of mechanical analyses. *J. Sed. Petrol.* **16**, 19–40.

Doeglas, D. J. (1971). Q_1 *Md* Q_3 indices showing grain size distribution on maps. *Geol. Mijnb.* **50**, 367–72.

Emery, K. O. (1938). Rapid method of mechanical analysis of sands. *J. Sed. Petrol.* **8**, 105–11.

Folger, D. W. (1972). Texture and organic carbon content of bottom sediments in some estuaries of the United States. *Geol. Soc. Amer.*, Mem. 133, 391–408.

Folk, R. L. (1966). A review of grain-size parameters. *Sedimentology*, **6**, 73–93.

Folk, R. L. (1974). *Petrology of sedimentary rocks.* Hemphill Publishing Co., Austin, Texas.

Folk, R. L. & Ward, W. C. (1957). Brazos River bar: a study in the significance of grain size parameters. *J. Sed. Petrol.* **27**, 3–26.

Ginsburg, R. N. & Lloyd, M. R. (1965). A manual piston coring device for use in shallow water. *J. Sed. Petrol.* **26**, 164–6.

Griffiths, J. C. (1967). *Scientific Method in the Analysis of Sediments.* McGraw-Hill, New York.

Guilcher, A. (1967). Origin of sediments in estuaries. In *Estuaries*, ed. G. H. Lauff, pp. 149–57. Pub. 83. Amer. Assoc. Adv. Sci., Washington.

Harbaugh, J. W. & Merriam, D. F. (1968). *Computer Applications in Stratigraphic Analysis*. John Wiley & Sons Inc., New York.

Hough, J. L. (1939). Bottom sampling apparatus. In *Recent Marine Sediments*, ed. P. D. Trask, pp. 631–64. Thos. Murby, London.

Hulsey, J. D. (1961). Relations of settling velocity of sand sized spheres and sample weight. *J. Sed. Petrol.* **31**, 101–12.

Inman, D. L. (1952). Measures for describing the size distribution of sediments. *J. Sed. Petrol.* **22**, 125–45.

Irani, R. R. & Callis, C. F. (1967). *Particle size measurement, interpretation and application*. John Wiley & Sons Inc., New York.

Kelland, N. C. (1975). In *Sea Surveying*, ed. F. A. Ingham. John Wiley, London.

Kirby, R. (1972). *The UCS vibrocorer 1972*. Unit of Coastal Sedimentation, Taunton, U.K., Report No. UCS/10/1972.

Klovan, J. E. (1966). The use of factor analysis in determining depositional environments from grain size distributions. *J. Sed. Petrol.* **36**, 115–26.

Klovan, J. E. (1975). R and Q-mode Factor Analysis. In *Concepts in geostatistics*, ed. R. B. McCammon. Springer-Verlag, Berlin.

Krumbein, W. C. (1934). Size frequency distributions of sediments. *J. Sed. Petrol.* **4**, 65–77.

Krumbein, W. C. (1938). Size frequency distributions of sediments and the normal phi curve. *J. Sed. Petrol.* **8**, 84–90.

Krumbein, W. C. (1939). Graphic presentation and statistical analysis of sedimentary data. In *Recent Marine Sediments*, ed. P. D. Trask, pp. 558–91. Thos. Murby, London.

Krumbein, W. C. & Pettijohn, F. J. (1938). *Manual of sedimentary petrography*. Plenum, New York.

Kuenen, Ph. H. (1968). Settling convection and grain size analysis. *J. Sed. Petrol.* **38**, 817–31.

Kullenberg, B. (1947). The piston core sampler, Sv. Hydr.-Biol. Komm. skr., 3, ser. Hydro. Bed. 1, H.2, Göteborg.

Lees, A., Buller, A. T. & Scott, J. (1969). Marine carbonate sedimentation processes, Connemara. *Reading Univ. Geol. Reports 1*.

McCammon, R. B. (1962). Efficiencies of percentile measures for describing the mean size and sorting of sedimentary particles. *J. Geol.* **70**, 453–65.

McCave, I. N. & Jarvis, J. (1973). Use of the Model T Coulter Counter in size analysis of fine to coarse sand. *Sedimentology*, **20**, 305–15.

McManus, D. A. (1965). A study of maximum load for small diameter sieves. *J. Sed. Petrol.* **35**, 792–6.

Martin, J. M. (1970). Utilisation des radioelements naturels en sedimentologie. *La Houille Blanche*, No. 8. 757–64.

Moss, A. J. (1962). The physical nature of common sandy and pebbly deposits, pt. 1. *Amer. J. Sci.* **260**, 337–73.

Moss, A. J. (1963). The physical nature of common sandy and pebbly deposits, pt. 2. *Amer. J. Sci.* **261**, 297–343.

Müller, G. (1967). *Methods in sedimentary petrology*. E. Schweizerbart'sche Verlagsbuchhandlung, Stuttgart.

Nelson, B. W. (1972). Biogeochemical variables in bottom sediments of the Rappahannock River Estuary. *Geol. Soc. Amer.* Mem. 133, 417–51.

Nichols, M. M. (1972). Sediments of the James River Estuary, Virginia. *Geo. Soc. Amer.* Mem. 133, 169–212.

Palmer, H. D. (1974). Estuarine sedimentation, Chesapeake Bay, Maryland, U.S.A. *Memoires de l'Institut de Géologie du Bassin d'Aquitaine*, No. 7, 215–24.

Passega, R. (1957). Texture as a characteristic of clastic deposition. *Bull. Amer. Assoc. Petrol. Geol.* **41**, 1952–84.

Passega, R. (1964). Grain size representation by CM patterns as a geological tool. *J. Sed. Petrol.* **34**, 830–47.

Passega, R. (1972). Sediment sorting related to basin mobility and environment. *Bull. Amer. Assoc. Petrol. Geol.* **56**, 2440–50.

Passega, R. (1974). Sorting variations between fluviatile and marine environments. *Mémoires de l'Institut de Géologie du Bassin d'Aquitaine*, No. 7, 381–5.

Peirce, T. J. & Williams, D. J. A. (1966). Experiments on certain aspects of sedimentation of estuarine muds. *Proc. Inst. civil Engrs.* **54**, 391–402.

Prokopovich, N. (1958). Settling tube for mechanical analysis by decantation. *J. Sed. Petrol.* **28**, 509–14.

Reineck, H. E. (1963). Der Kastengreifer. *Natur u. Museum*, **93**, 102–8.

Reinson, G. E. (1975). Geochemistry of muds from a shallow restricted estuary, Australia. *Mar. Geol.* **19**, 297–314.

Richardson, J. F. & Zaki, W. N. (1954). Sedimentation and fluidisation. *Trans. Inst. Chem. Eng.* **32**, 35–51.

Ryan, J. J. & Goodell, H. G. (1972). Marine geology and estuarine history of Mobile Bay, Alabama. Part 1. Contemporary Sediments. *Geol. Soc. Amer.* Mem. 133, 517–54.

Sagoe, K-M. O. & Visher, G. S. (1977). Population breaks in grain-size distributions of sand. A theoretical model. *J. Sed. Petrol.* **47**, 285–310.

Sanford, R. B. & Swift, D. J. P. (1971). Comparison of sieving and settling techniques for size analysis, using a Benthos rapid sediment analyser. *Sedimentology*, **7**, 257–64.

Schlee, J. (1966). A modified Woods Hole rapid sediment analyser. *J. Sed. Petrol.* **36**, 404–13.

Sengupta, S. & Veenstraa, H. J. (1968). On sieving and settling techniques for sand analysis. *Sedimentology*, **11**, 83–98.

Shepard, F. P. (1973). *Submarine Geology* (3rd edition). Harper & Row, New York.

Silverman, M. & Whaley, R. C. (1952). Adaption of the piston coring device to shallow water sampling. *J. Sed. Petrol.* **22**, 11–16.

Trask, P. D. (1930). Mechanical analysis of sediments by centrifuge. *Econ. Geol.* **25**, 581–99.

Trask, P. D. (1932). *Origin and environment of source sediments of petroleum.* Gulf Publ., Houston, Texas.

Urien, C. M. (1972). Rio de la Plata Estuary Environments. *Geol. Soc. Amer.* Mem. 133, 213–34.

Van Veen, J. (1936). *Onderzoekingen in de Hoofden.* Algemeene Landsdukkerij, 'S-Gravenhage.

Visher, G. S. (1969). Grain size distributions and depositional processes. *J. Sed. Petrol.* **39**, 1074–106.

Visher, G. S. & Howard, J. D. (1974). Dynamic relationship between hydraulics and sedimentation in the Altahama estuary. *J. Sed. Petrol.* **44**, 502–21.

Wentworth, C. K. (1922). A scale of grade and class terms for clastic sediments. *J. Geol.* **30**, 377–92.

Williams, D. J. A. & McManus, J. (1975). Tidal dynamics and sediment distribution in the Tay Estuary. *Proc. Int. Sedim. Congr.* Nice. Thème 3, 121–9.

Wilson, J. V. (1967). Palaeoecological studies on shell beds and associated sediments in the Solway Firth. *Scottish J. Geol.* **3**, 329–71.

Zeigler, J. M. & Gill, B. (1959). Tables and graphs for the settling velocity of quartz in water, above the range of Stokes' Law. Rep. 59–36. Woods Hole Oceanographic Institute.

6

Suspended sediment

I. N. McCAVE

To the casual obvserver, suspended material is one of the most obvious features of estuaries. They tend to be muddy; muddier than the rivers flowing into them or the sea beyond. In fact, most estuarine circulation systems lead to the entrapment of suspended sediment, making the study of this material of prime importance as aspects of both estuarine behaviour and pollutant dynamics. The location of many industries along the banks of estuaries makes the latter aspect very pressing. This chapter deals with the determination of suspended material concentration, properties and interpretation of results. The terms suspended material, suspended sediment, suspended matter, suspensate (USA) and Seston (German usage) all refer to total solids held in suspension in water, though operationally what is retained on a filter is determined. These terms will be used as synonymous and interchangeable.

Determination of concentration

Water sampling and filtration

Water sampling and samplers

The simplest method for the determination of suspended sediment concentration is to take a known volume of water, filter it and determine the weight of material retained on the filter. The amount of suspended material needed for determination of concentration depends on the type of filter used, but most need less than 50 mg. Most estuaries have concentrations of more than 10 mg/l and consequently large samples are not necessary, samples of one litre (exceptionally up to 5 l) being adequate.

Water bottles. There are many types and makes of water bottle. The two principal types are represented by the NIO bottle (from the formerly named British National Institute of Oceanography) with capacities of 1.25–7 l and the Van Dorn bottle made with capacities of 1–10 l and sometimes more. The NIO bottle (Fig. 6.1) is an open PVC tube with hemispherical bungs on spring

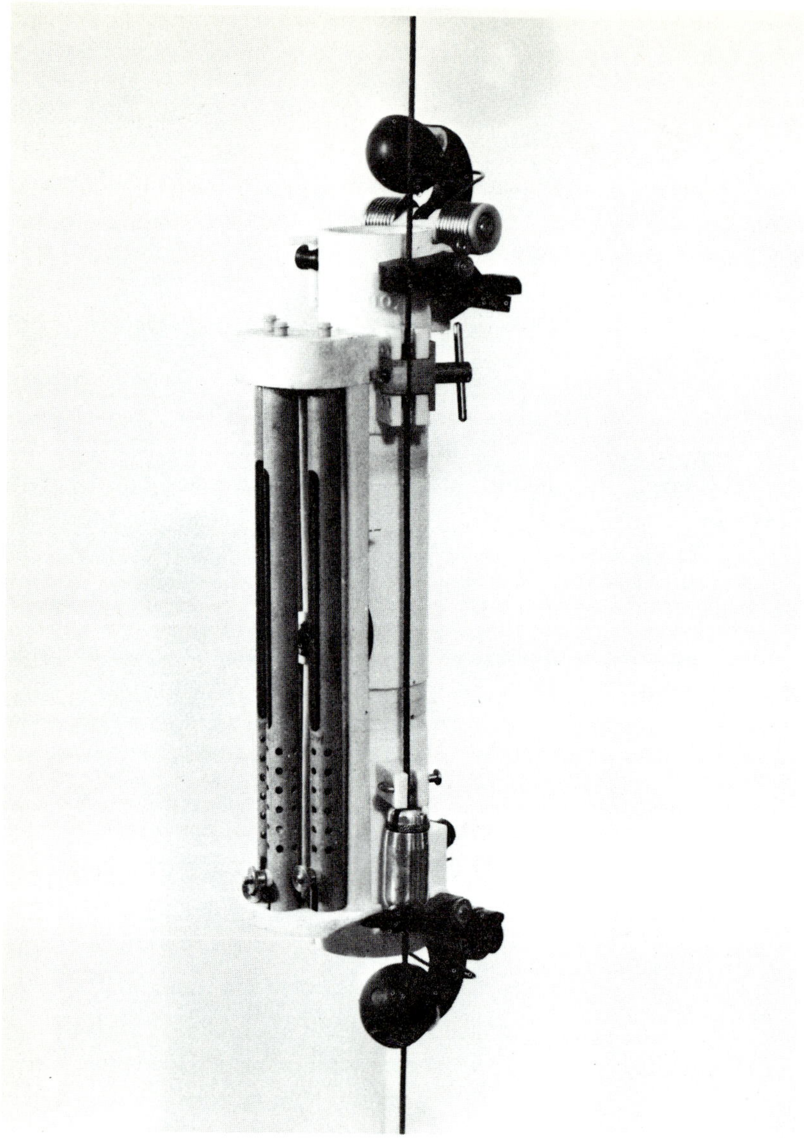

Fig. 6.1. NIO Bottle.

loaded arms to close each end. As the diameter of the openings is the same as that of the body of the bottle, it has good flushing characteristics. Samples such as the older Nansen bottles, which do not have this characteristic, should not be used for suspended material work. The Van Dorn bottle is an open tube normally made of PVC, with a rubber spring through the inside of the bottle and connected to end stoppers. To set the bottle these are pulled out and connected to a release mechanism actuated by messenger. Currently the most commonly used bottle of this type is the Niskin bottle which is available in sizes from 1 to 30 l. If trace-element work is to be undertaken, the internal rubber spring of these bottles should be replaced. Teflon coated steel springs have been used as a substitute, but Bewers, Macaulay & Sundby (1974) find these tend to abrade and peel and recommend a PTFE-sheathed shock cord.

Traditionally, water bottles have been closed by dropping a brass messenger down the hydrographic wire. The bottles are attached to the wire at preset positions and after lowering to the desired set of depths are triggered by messenger. If particular points in the water column identified by temperature/salinity or light measurements are to be sampled, then these casts are usually made first. However there is now a Niskin 'rosette', which holds 12 bottles and contains a mechanism which allows them to be triggered one at a time from the surface. Temperature/salinity and light measuring devices can also be mounted on the rosette allowing the water column to be sampled at points of interest. Needless to say the cost is high ($>$ £10^3). Such a device would encounter considerable problems if operated in a shallow, fast flowing estuary and the traditional method is probably still the best.

There are many minor variants of the bottle types described here and it is a quite straightforward job to make one's own (e.g. Duursma, 1967). The important points to keep in mind are: suitable size, tight closure, ease of handling and lack of contamination. If metal analyses are to be made, a good quality rust-free hydrographic wire should be used as the messenger will cause spallation of rust particles which may be collected in the bottle.

Pumps and traps. Most of the material in suspension in an estuary is fine-grained, and has a quartz-equivalent settling diameter less than ~ 50 μm (settling velocity < 0.2 cm s^{-1}) even in aggregates. However many estuaries have sand beds and close to the bed there is an appreciable concentration of suspended silt and sand. In order to sample this both pumps and traps have been used. Crickmore & Aked (1975) described a system with a shipboard pump and a hose with a movable nozzle mounted in a bed frame. Postma (1965) also used a pump with a hose lowered over the side and passed 200 l of water through a 35 μm mesh plankton net. This simple system provides sufficient material for grain-size analysis but is not easy to use in fast flowing water and cannot be positioned accurately close to the bed. For that situation, there are sediment traps which extract material from suspension. One such is the Delft Bottle. This has a nozzle pointing into the flow which opens into an expansion chamber shaped like a bottle and with holes in its rear plate (Fig. 6.2). Water entering the nozzle slows down in the expanded volume of

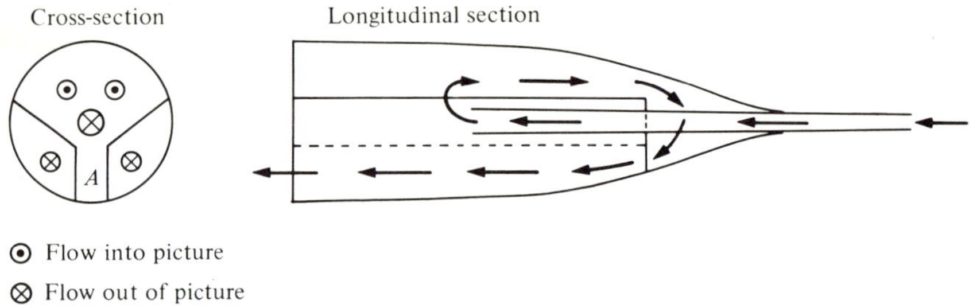

Fig. 6.2. Delft Bottle: schematic diagram showing flow pattern. Most trapped sand accumulates at point *A*.

the bottle, drops its load of suspended sand in the bottle and passes out of the holes in the rear. It catches material > 50 μm in diameter and in 15–30 min sufficient can be obtained to perform a size analysis. In order to determine concentration a measurement of flow speed at the nozzle height of the bottle is also required.

In estuaries with high water speeds and a high proportion of sand in the suspended material, conventional water bottles obtain an anomalously low sample. This is probably because some of the sand tends to settle out of the bottle before it is triggered. The best solution to this problem is to use a river suspended sediment sampler. Various models in common use in the USA are shown in Vanoni (1975, pp. 325–42). They all have a nozzle pointing into the flow and are rather heavy (25 kg and up).

Harvesting. At times one wishes to obtain a large quantity of suspended material for chemical or mineralogical analysis. In this case it may be necessary to obtain hundreds of litres of water and extract sediment from it. This may be done by large diameter filters, filter candles or by continuous flow centrifuge (see below for details). The samples may be obtained by pump and hose (best), repeated use of water bottles (tedious and time consuming) or by bucket (worst but probably most common). The bucket method may yield samples biased by low-density material concentrated in the surface layer.

Filtration and filters

Filtration systems. Filtration may be carried out on board ship or in the laboratory. The former procedure removes the necessity of transporting many bulky water samples. Filtration may be carried out either by suction (the commonest method) or under pressure. When working with a suction system it is usual to run a number of vacuum flasks off one pump (Fig. 6.3). In this system open filter holders are mounted on each flask. Large, easy-to-screw-down clamps are used on all lines to flasks so that they may be isolated. This allows the pump to be switched off much of the time; a necessity as continuous operation over tidal cycles, 12½ or 25 h, can lead to overheating problems. A water-trap is included between the vacuum line and the pump as salt water is very bad for vacuum pumps! Care should be taken in setting up a leak-free system. The filter holders should be kept clean by a cover of aluminium foil or by removal to a covered container when not in use. High vacuum should *not* be used for filtration, 35 cm Hg of vacuum is sufficient. High vacuum pulls particles into the filter, making it clog sooner.

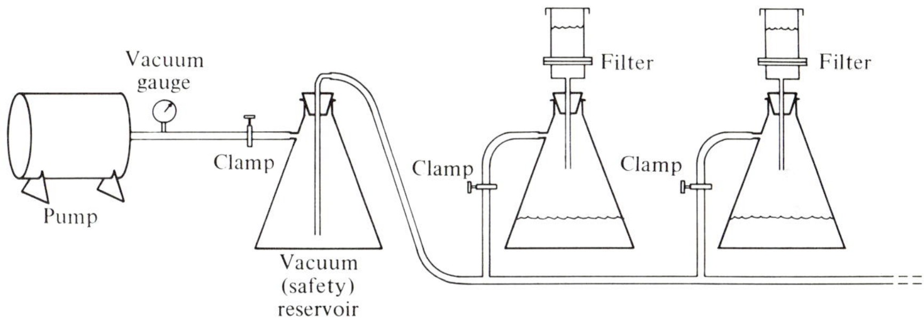

Fig. 6.3. Filtration system using simple suction.

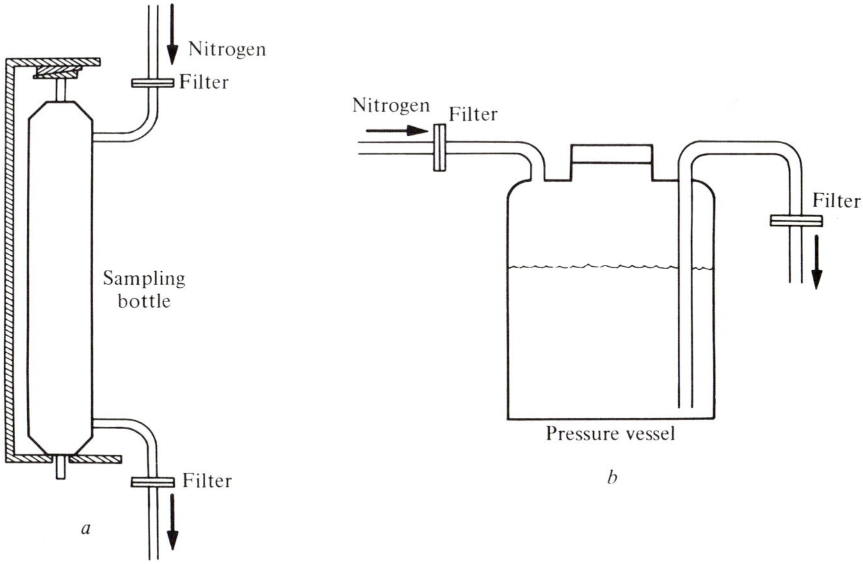

Fig. 6.4. Pressure filtration. *a*. The sampling bottle (after Sundby, 1974). *b*. Transfer of the sample to another container.

Pressure filtration may be performed direct from the sampling bottle (Fig. 6.4) or after transferring the sample to a pressure vessel. The former method avoids contamination. Filtered nitrogen at pressures up to about $\frac{1}{2}$ at. is used. A modification of this method employs a small back pressure at the outlet end so that the filter has liquid on both sides of it rather than air on the exhaust side. This flow-through pressure filtration system improves the preservation of delicate particles on the filter. In general in-line systems such as that in Fig. 6.4, whether pressure or suction driven, are to be preferred because they minimise contamination.

Procedures. Filters should be pre-weighed and kept in individual numbered petri dishes. They should be handled only at the edges using flat-bladed tweezers. Following filtration the filters should be washed with filtered distilled water (remember, if you are to analyse the filtered water to remove it *before* washing through with distilled water). Opinion varies on washing procedures but a number of applications of small volumes of water (say 5×10 ml on a 47 mm diameter filter) are better than one big one. With many types of filters it is necessary to run blanks or at least to have established beforehand the magnitude of possible weight changes in the filters themselves. Blanks may be run at the same time as filtering. Two filters are put in the same holder and the lower one (blank) is subsequently treated in exactly the same way as its pair. Alternatively, filtered estuary water from one sample may be run through a filter to establish blank values. Blank procedures will be dealt with under filter types.

Where the sample contains appreciable amounts of suspended sand, it is difficult to obtain a representative sub-sample for filtration. In this case it is best to pass the whole volume of the sample through a 63 μm sieve, remove, dry and weigh that fraction separately and filter an aliquot of the remaining suspension. The weights are then combined using the appropriate factor to account for the splitting, and the concentration of the whole is produced. The concentration of sand (> 63 μm) is also a useful parameter on its own.

One problem which must be guarded against is the particles settling in the sampling bottle. In the case of pressure filtration (Fig. 6.4) some of the larger particles may settle below the lower outlet. The bottle thus needs to be shaken to ensure filtration of a homogeneous and representative suspension. Also there may be so much material in suspension that the filters clog before all the water is filtered. In this case it is particularly important to ensure that what *is* filtered is a homogenised sub-sample of the whole. Before taking a sub-sample for filtration it may be necessary to withdraw a little water before shaking the bottle and then replacing the water. In particularly turbid waters one should not blithely set out to filter a litre of suspension only to find that after 150 ml the filter is clogged and that there is another 100 ml of water in the filter holder and distilled water washes have to be filtered as well! The filtration procedure needs planning, knowing filter loadings attainable in practice, and having some idea of the conditions likely to be encountered.

Drying. A number of authors suggest oven drying and desiccation of filters. I find that it is impossible to obtain an accurate weight by this method because on removal from the desiccator to the balance the filter is acquiring moisture from the atmosphere. Accurate weighing cannot be done in a hurry so the filter needs to equilibrate with the atmosphere. The best way of drying the filters is in the air, in individual petri dishes with the covers over the dish but not firmly on to allow circulation of air. Ideally this drying and the weighing should be carried out in an atmosphere of controlled humidity. A laminar-flow air cabinet over the balance is best as it can also be used as a drying environment for the filters. Blanks are run to check on weight changes as detailed below. Any form of oven drying before or after filtration is not recommended.

Weighing. For most work a balance reading to 10^{-5} g will be sufficient. Only in situations with heavily loaded filters and/or where order of magnitude results are satisfactory will a 10^{-4} g balance suffice. Many of the filters are electrostatically charged and the balance should contain an ionising source to allow the charge to be dispersed. Generally, α emitters (e.g. ^{210}Po or ^{241}Am) are used. All the usual precautions for precision weighing should be observed. They include: checking the zero, checking the reading with some repeatedly weighed object, and allowing 10 min or so to elapse while doing trial weighings before doing samples. Thus a 'quick' weighing of a few filters cannot be done with the same precision as a series. The standard deviation of a repeatedly weighed filter on my 10^{-5} g balance is 0.025 mg. Sundby (1974) obtains the same value and points out that the standard deviation of the difference between two filter weights becomes $0.025\sqrt{2}$ or 0.035 mg. For a 'quick' weighing this might well be 0.1–0.2 mg on a 10^{-5} g balance. Additional errors are incurred in estimating the volume filtered. Where this is very small, a few hundred millilitres in turbid estuaries, a measurement of volume to better than 5 ml is needed.

Filters. The principal types of filters in current use are cellulose ester membranes (CEM), track-etched membranes (TEM), metal membranes (MM), glass microfibre (GM) and, for harvesting, ceramic candles. Filters are made in a great variety of sizes and nominal pore diameters. Those most commonly used for marine work are 47 mm diameter with a nominal pore diameter of 0.4 μm. Pore sizes of 0.8 μm are sometimes used, and larger filter diameters, 14 cm and 29 cm, are also used on occasion, but handling and storage are more difficult.

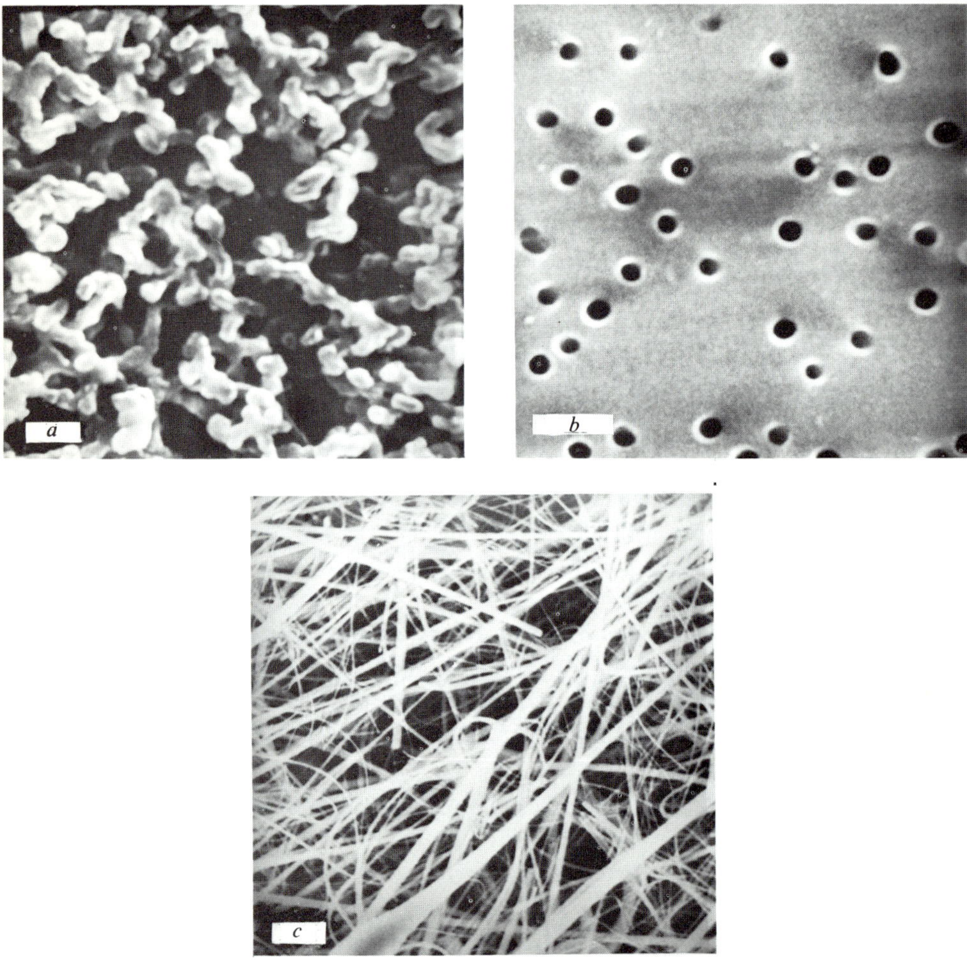

Fig. 6.5. Scanning electron micrographs. *a*. A cellulose ester membrane. Scale bar 1 μm. *b*. A perforated polycarbonate membrane. Scale bar 1 μm. *c*. A glass fibre filter. Scale bar 10 μm.

Cellulose ester membrane filters are microporous discs with a narrow range of pore sizes about the nominal size (e.g. 0.45 ± 0.02 μm) (Fig. 6.5*a*). The principal makes are Millipore, Sartorious and the less expensive (in UK) Oxoid. They are about 150 μm thick and a 47 mm filter weighs about 100 mg. Their advantages are that a moderate sediment loading of 20–40 mg can be obtained (1.5 to 3 mg cm^{-2}) at a reasonable initial flow rate (64 ml cm^{-2} min^{-1} distilled water at 1 at. differential pressure). In addition they can be rendered transparent by a few drops of cedar oil (Refractive Index, 1.51) for microscopic examination of the filtrate. The disadvantages are that they are hygroscopic so their weight fluctuates with humidity, and that they contain soluble material which is washed out. A 47 mm filter loses anywhere from 0.2 to 0.5 mg in this way. Various pre-soaking techniques have been suggested, but these tend to make the filters brittle and render them liable to early clogging. Pre-soaking and oven drying (Banse, Falls & Hobson, 1963; Eaton, Likens & Bormann, 1969) are *not* recommended. Blanks are thus absolutely necessary. The most accurate work use paired filters for every sample (Eaton *et al.*, 1969). Less rigorous work may use one or two blanks in ten samples. Average or batch blank corrections are then applied. Even the paired filter technique is of doubtful accuracy as the loaded filters need not have had exactly the same loss of soluble material as their blank pair. Most of the sediment can be removed from the filter by back flushing while it is still wet. However once it has been dried it is difficult to resuspend the sediment. Material can however be released from the filter by dissolving the filter in acetone. Relative advantages and disadvantages of cellulose (Millipore) and track-etched (Nuclepore) filters are given by Biscaye & Eittreim (1974, pp. 235–7).

Developed in the last few years, the track etched membrane is a 10 μm thick polycarbonate plastic film irradiated in a nuclear reactor to leave tracks which are then etched into cylindrical pores (Fig. 6.5*b*). At present Nuclepore is the only make. Variation in etching time yields pores of differing diameter. A 47 mm filter weighs about 18 mg. The advantages of these filters are their low weight, lack of detectable weight loss when washed and relative insensitivity to changes in humidity. These properties mean that with careful weighing and humidity control in the laboratory it is not necessary to use blanks. Also the suspensate collected on the filter can be almost totally removed ($\sim 97\%$, 0.05 mg remaining) by ultrasonic treatment. Disadvantages are: only low loading is possible, not much more than 5 mg on a 47 mm filter (0.4 mg cm^{-2}) can easily be achieved, and the flow rate falls off rapidly. The low loading places a premium upon accurate weighing procedures. As the cellulose ester filter tends to clog slowly it retains more of the particles finer than the nominal diameter than does a Nuclepore filter. Furthermore, the Nuclepore filter cannot be rendered transparent by oils for microscopic analysis of the filtrate.

Silver filters do not now appear to be widely used for either gravimetric analysis or harvesting. Standard sizes and pore diameters are available from the Selas Corporation but they are not cheap. Their advantages are that they have good flow rate, achieve moderate loading and can be easily backflushed. They can be cleaned of dried sediment making them re-usable, and they make good mounts for X-ray diffraction studies. However they tend to lose weight by dissolution during filtration of seawater (Gaudette & Anderson, 1970). While these filters are not best for field use (the X-ray diffraction traces from untreated suspended material are poor, e.g. Manheim, Hathaway & Uchupi, 1972), they are useful in the laboratory in diffraction work as a substrate for suspended material cleaned of its organic matter.

Glass microfibre filters are discs of compressed glass fibre forming a 'depth filter' rather like a piece of felt (Fig. 6.5*c*). Familiar brands are Whatman and Gelman. Whatman GF/C filters of 47 mm diameter are 260 μm thick, weigh about 100 mg and have a nominal minimum particle retention size of 1.2 μm. They have a high flow rate and take a high load. They can be pre-combusted to remove all traces of organic carbon and this property makes them an integral part of one method for organic carbon determination. They tend to shed fibres so are not completely weight-stable. Although a large amount of sediment can be extracted, it is difficult to remove, as much may be embedded in the filter. Samples filtered for organic carbon analysis should not be rinsed with distilled water as this will rupture membranes resulting in some loss of organic matter.

Ceramic filter candles are hollow tubes of microporous porcelain. Nominal retentions of 0.06–25 μm can be obtained. Values of 0.3–0.7 μm are best for normal laboratory use. The candle is attached to a vacuum line and immersed in the suspension. The water is sucked through the walls leaving behind the mud. Many candles can be run off one line (which must have a large volume water-trap). This is a good way of concentrating material from large volumes of sample. Material embedded in the candle is released by applying back pressure.

Continuous flow centrifuge. Another method of separating sediment from large volumes of water is to pass the water through a continuous flow centrifuge. This is a bucket rotating at high speed with suspension fed into the bottom and clear fluid skimmed off the top (Fig. 6.6). In this set-up the time taken to deposit a particle, of diameter d and density ρ_s, from a fluid of dynamic viscosity μ and density ρ in a rotor spinning at angular velocity ω, is

$$T = \frac{\ln (r_2/r_1)\, 18\,\mu}{(\rho_s - \rho)\,\omega^2 d^{\,2}}. \tag{6.1}$$

If V is the volume of fluid in the rotor when spinning then the flow rate required to trap particles of size d is $Q = V/T$. For example with $\rho_s = 2.5$ g

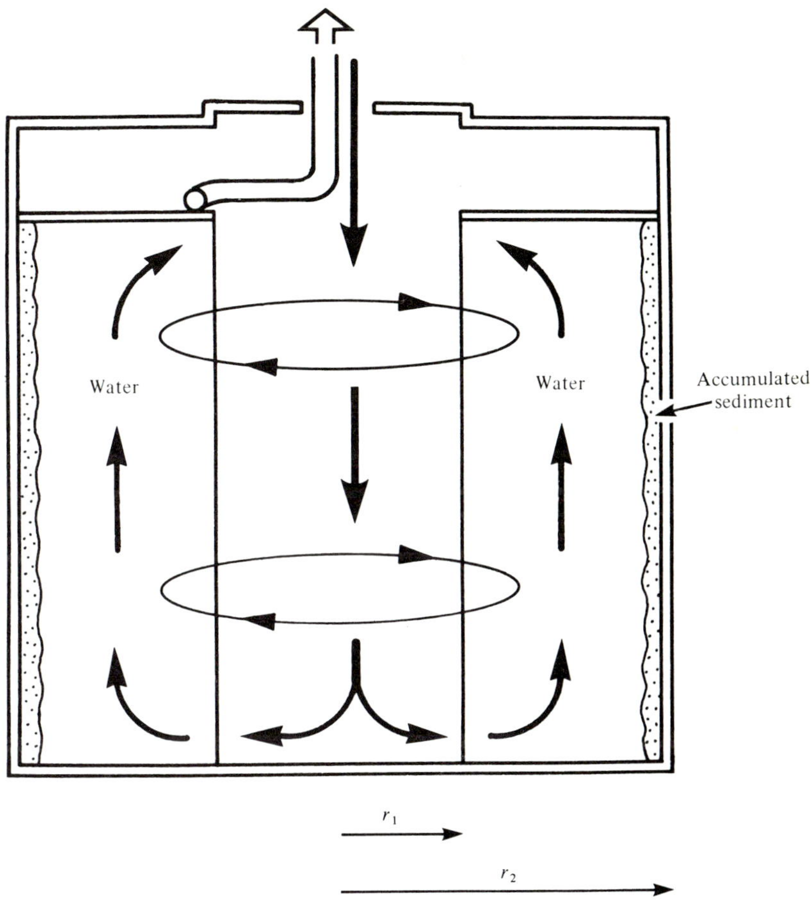

Fig. 6.6. Flow system and trapping of sediment in the rotor of a continuous flow centrifuge. The radii r_1 and r_2 are referred to in the text.

cm^{-3}, $\rho = 1$ g cm^{-3}, $\mu = 0.01$ poise, $d = 10^{-4}$ cm (1 μm), rotation rate = 18000 rpm ($\omega = 2\pi$ 18000/60), $r_2 = 5$ cm, $r_1 = 3$ cm and $V = 200$ ml then $T = 1.73$ s and the flow rate is 116 ml s^{-1}. However the relation depends on d^2 and for smaller sizes the flow rate becomes small, 29 ml s^{-1} for 0.5 m particles and 1.16 ml s^{-1} for 0.1 μm particles. Another variable which poses an acute problem is the bulk density of suspended particles; ρ_s is generally lower than the mineral density of some of their constituents. Much suspended material is in the form of flocs or aggregates, frequently with organic matter. A more realistic density might be 1.2 g cm^{-3} giving $(\rho_s - \rho) = 0.2$ rather than 1.5 and decreasing the flow rates given above to 15, 3.9 and 0.15 ml s^{-1}. This gives very long running times and it may be concluded that the filter candle is simpler, cheaper and more effective.

Conclusion. For precision work determination of low concentration is best done with Nuclepore filters. Because of the small volumes of water required (maybe only 50 ml before a Nuclepore clogs) a pipette system should be available for measurement of small sample volume. In most estuaries where concentrations are moderate to high (> 10 mg/l), cellulose ester membranes (Sartorious, Oxoid, Millipore) are best. The loadings are sufficiently high so the weight-loss can be corrected for without great error; they are easier to handle and can be used under the microscope with oils. For collection of larger amounts of sediment 14 cm diameter cellulose ester membranes can accommodate 200 to 600 mg of sediment. For even larger amounts filter candles are suitable. Glass fibre filters are of particular importance for organic carbon work and silver membranes are of use as X-ray diffraction substrates.

Optical methods

For many years it has been common to examine the optical properties of a suspension as a guide to the amount and characteristics of the suspensate. Workers must not lose sight of the fact that they are making optical measurements when they shine lights at detectors, *not* measurements of suspended material content. The principal optical measurements are of beam attenuation (or transmittance) and scatterance. In the former the light beam is shone directly at the sensor, whereas in the latter the sensor measures the light scattered (reflected and refracted) by suspended particles. The measuring instruments are the beam transmittance meter, or transmissometer, and scatterance meter, or nephelometer, and the Secchi disc. Both transmittance and scatterance are functions of not only the number, but also the size, index of refraction and shape of the particles. Thus different suspensions of the same concentration will have different optical properties. It is therefore undesirable for scientific work to calibrate optical sensors in terms of artificial turbidity standards, e.g. the polymer formazin (Eden, 1965; American Public Health Association, 1971). One undesirable feature of such calibration is that the units are often equated with mg l^{-1} ($\sim$ ppm by weight) and are reported as such. Nephelometers are sensitive to very low concentrations and do not find much use in estuaries so more attention will be given here to transmissometers. Greater detail on optics and instrumentation will be found in Jerlov (1976), Jerlov & Nielsen (1974), Williams (1970) and Gibbs (1974).

Beam transmittance

The transmittance (T) of sea water is the ratio of the transmitted (F_L) to the incident (F_O) radiant flux (Fig. 6.7). The relation $T = (F_L/F_O) = e^{-\alpha L}$ defines the attenuation coefficient α measured over a path length L. In practice the irradiance E ($= dF/dA$) on a surface area A at the source and after travelling

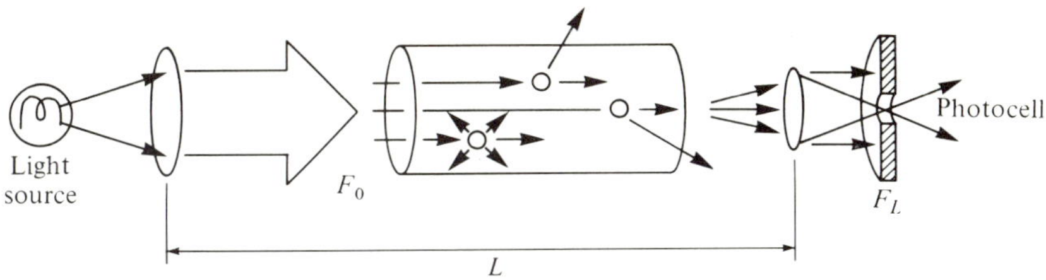

Fig. 6.7. Schematic diagram of a transmissometer. The incident light flux F_O is scattered and absorbed so a smaller flux F_L is received at the end of the light path L. From *Optical Properties of the Sea* by J. Williams. Copyright © 1970, US Naval Institute.

L is measured using photo-electric cells. A very narrow beam ($\sim 1\ \text{cm}^2$) over a path of 1 m is suitable to ensure that the width/length is small. A reading of the transmittance T is obtained and from this the coefficient α is derived. Estuaries have such a range of turbidities that an instrument with a variable path L is desirable. Also available are double beam transmissometers in which the source beam is split and sent along two paths of length L_1 and L_2. The transmittances are $T_1 = e^{-\alpha L_1}$ and $T_2 = e^{-\alpha L_2}$ and their ratio

$$T_2/T_1 = e^{-\alpha(L_2 - L_1)}. \tag{6.2}$$

If the set-up is such that $L_2 = 2L_1$ then

$$T_2/T_1 = e^{-\alpha L_1} \tag{6.3}$$

so the attenuation coefficient can be obtained from the ratio of the two values of transmittance.

Parallel light suffers a decrease in intensity on passing through sea water because of absorption by water and particles. Absorption by particles is a strong function of wave length, and organic particles have strong absorption in the blue at $\sim 0.45\ \mu\text{m}$ with another peak in the red at $0.67\ \mu\text{m}$. This is partly responsible for the variable ratio between α or T and suspended material concentration (Fig. 6.8). The better types of transmissometer have sets of filters which can be changed *in situ* allowing observations to be made with monochromatic light in different parts of the spectrum.

There are various types of 'turbidity meter' on the market. These do not allow one to obtain the basic optical parameters as the beam width/length is not small and some scattering occurs. They should be calibrated with reference to samples of the natural mineralogy and it is most important to check that this calibration does not drift during use. A double-beam device of this type is described with a summary of the principle by Thorn (1975*a*). The double-beam system eliminates the problem of fouling of the lenses *assuming* that all are equally contaminated. The device does not allow the

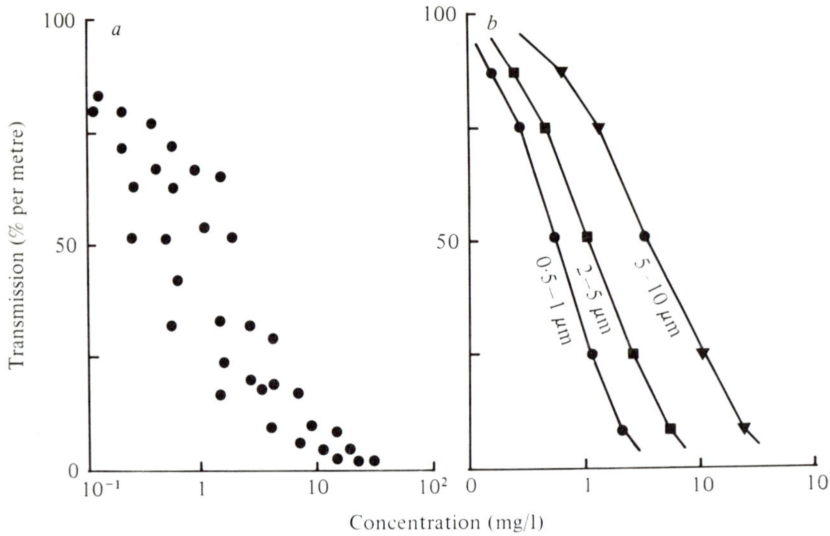

Fig. 6.8. Relationship between transmittance and suspended material content (*a*) and the same for specific sizes of material (*b*). From Gibbs (1974).

optical parameter α to be obtained and it is usually calibrated with respect to Formazin. Both this type of reading and α can be calibrated by filtering samples of water from known points in the water column for which a light attenuation trace is available. If any mention is to be made of *concentration* based on light profiles whether by turbidity meter or α-meter this form of calibration is *essential*. Of course the scatter is likely to be great (Fig. 6.8). Values given by Jones & Wallis (1956) for suspensions of Thames mud and for the English Channel and Thames Estuary are $\alpha = 0.51C$ and $\alpha = 0.25C$ where C is the suspended matter concentration in mg/l. A less suitable value α_R is used by Joseph (1953) and Otto (1966). This is given by

$$\alpha_R = \frac{1}{L}\ln\frac{F_{L\,\max}}{F_L}, \tag{6.4}$$

where $F_{L\max}$ is the maximum value of F_L found in a series of observations. Thus the measurement is relative to the clearest water encountered (α_o), but this may be quite turbid. Thus

$$\alpha_R + \alpha_o = \alpha$$

Otto's (1966) measurements in the Wadden Sea gave two groups of values with α_R proportional to $0.06C$ and $0.13C$, distinctly less than the absolute values of Jones & Willis.

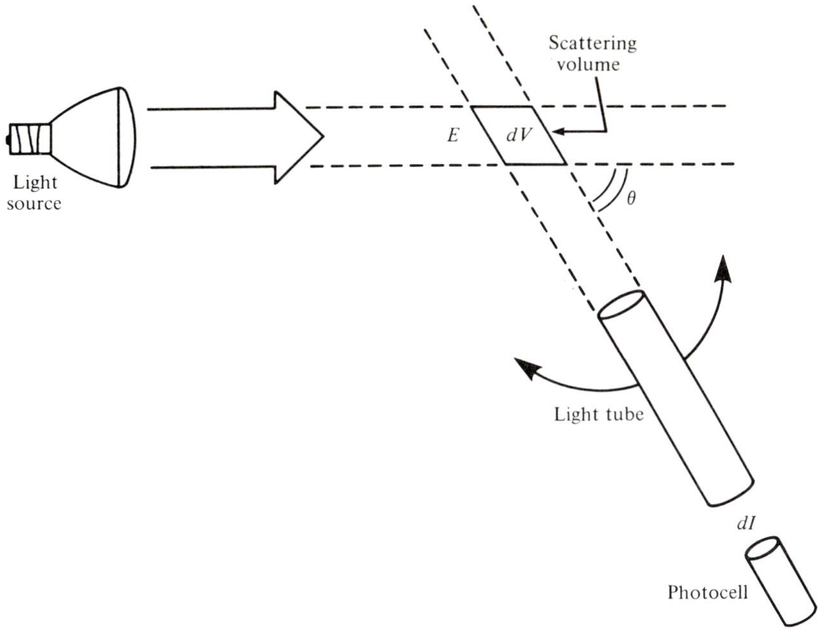

Fig. 6.9. Schematic diagram of a nephelometer. θ is the scattering angle. Scattered light is collected and passes through a black tube to the measuring device (photocell). From *Optical Properties of the Sea* by J. Williams. Copyright © 1970, US Naval Institute.

Scattering measurement (Nephelometry)

Scattering measurements are more complicated than transmittance and calibration is difficult. Such measurements have not been reported from the relatively turbid waters of estuaries though Pak, Beardsley & Park (1970) came close. The radiant intensity (I) of scattered light from a beam is a function of the angle to the beam at which the measurement is made θ, the incident irradiance E and the volume irradiated V (Fig. 6.9). The volume scattering function is then

$$\beta(\theta) = \frac{\mathrm{d}I}{E\,\mathrm{d}V}. \tag{6.5}$$

The units of radiant intensity are watts steradian^{-1} (the steradian is the measure of solid angle), of irradiance watts m^{-2} and $V = \mathrm{m}^3$, so $\beta(\theta)$ has units of (m ster)$^{-1}$. Total scattering b (units of m^{-1}) is given by

$$b = \int_{4\pi} \beta(\theta)\,\mathrm{d}\omega = 2\pi\int_0^{\pi} \beta(\theta)\sin\theta\,\mathrm{d}\theta. \tag{6.6}$$

It has been observed by a number of authors that b is proportional to β (45°)

and thus that an estimate of total scattering (argued to be linearly related to particle concentration) can be obtained from one measurement, $\beta(45°)$, (e.g. Beardsley *et al.*, 1970). However the ratio $\beta(45°)/b$ varies between water masses (Smith, Austin & Petzold, 1974).

The scatterance varies strongly with θ, being high at small forward angles. This is not easy to measure with conventional light sources but the advent of lasers is changing that. A forward-scattering laser nephelometer used in shelf suspended sediment work is described by Sachs in Meade *et al.* (1975). Such instruments may turn out to be very useful in estuaries.

Secchi disc

The Secchi disc is a white disc of 20–30 cm diameter which is lowered into the water to a depth D at which it just disappears from view. The factors which determine D are many but certainly include α and the relative-irradiance loss coefficient, the relative amount of upwelling light, and the surface reflectance. The attenuation and loss coefficients are partly controlled by the amount, nature and size of the suspended material. Detailed discussions of Secchi discs measurements and their meaning are given by Postma (1961*a*), Otto (1966), Tyler (1968) and Williams (1970, p. 72). The calibration of the disc in terms of suspended material concentration can only be done by taking simultaneous water samples for gravimetric analysis. Having some rough calibration curve, the disc depth does provide a quick estimate of conditions and this is its chief value. However only surface values are obtained.

Gamma ray measurements

Particularly concentrated suspensions may have sufficiently high natural gamma activity to allow relation of this to concentration and thus its use as an *in situ* method for obtaining suspended material profiles. A complementary use of artificial gamma rays is as a source and detector array to determine the attenuation of the beam. This gives a measure of the density of the suspension.

Natural radioactivity

The measurement of natural gamma activity was carried out in the Gironde Estuary by Martin (Martin & Chesselet, 1969; Martin, Meybeck & Heuzel, 1970). Measurements were made through the tidal cycle at a number of fixed sites along the estuary at 0.8 and 3 m above the bed. In this estuary suspended sediment concentrations are high. Corrections have to be applied for ^{40}K in the water, cosmic radiation which varies with depth, and the influence of the radioactivity of the bed and the equipment. It was found that the corrected count rate from the detector N_t was related to suspended material content

C g l^{-1} by $N_t = aC^b$ with $a = 0.18$ and $b = 0.8$. C was determined by millipore filtration. Concentrations over 100 g l^{-1} were measured by scintillation counter.

There are few examples of this technique as it requires concentrations $C > 0.5$ g l^{-1}. However, it may prove useful in the study of fluid mud.

Gamma-ray densitometry

This method has been applied for a few years to the determination of sediment density ρ in cores while still in their liner tubes (Preiss, 1968; Whitmarsh, 1971). The basic relationship is $\rho = (1/\mu L)\ln(I_o/I)$ where μ is the Compton mass attenuation coefficient, L the path length through sediment and I and I_o are the gamma ray intensities with and without the core in the path of the beam. Recently this has been applied to dense suspensions (1.15–1.35 g cm^{-3}

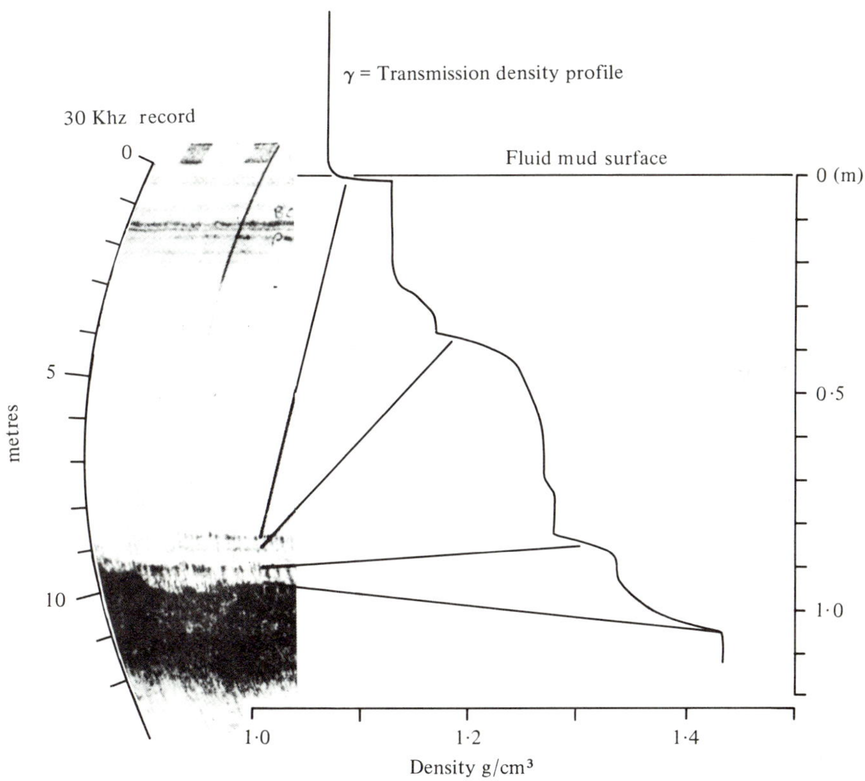

Fig. 6.10. Echo sounder record (left) showing 'false echoes' indicating presence of layered fluid mud. γ-ray transmission density profile is shown on the right displaying a number of steps from 1.1 to 1.4 g cm^{-3}. (From Kirby & Parker, 1974.)

or 250–575 g l^{-1}) by Kirby & Parker (1974). They give no details of their instrumentation save that the path length is 15 cm and the precision is $\pm 2.5\%$. However they do show excellent correlation between density steps within fluid mud pools and echo sounder records taken at the same time, showing reflecting layers within the fluid mud (Fig. 6.10). Fluid mud is assuming an increasing importance in estuarine studies because very large quantities of suspended material dominating the estuarine budget are temporarily stored in the fluid mud pools. Not all estuaries have such fluid-mud accumulations but in those that do, gamma-ray densitometry will be a valuable aid.

Remote sensing

A considerable and rapidly increasing number of sophisticated instruments orbit the earth and survey many features of its surface. Of particular importance to suspended sediment monitoring are the scanners in the visible light range. The satellites of principal interest are the ERTS 1 (Earth Resources Technology Satellite) launched in 1972, now supplemented by LANDSAT 2 launched in 1974, and Nimbus-G to be launched in 1978. Their characteristics are reviewed by Apel (1976).

ERTS-1/LANDSAT-2

These carry the multispectral scanner (MSS) with seven wave-length bands of which bands 4 (0.5–0.6 μm) green, 5 (0.6–0.7 μm) red and 6 (0.7–0.8 μm) near infra-red are of most interest. The satellite senses the radiance from the earth surface (watts m^{-2} steradian^{-1}) in the given wave-length bands. Suspended sediment shows up quite well in band 5, and the radiance emanates from the upper ~2 m of the water column. For band 6 the source is only the top 10 cm. The theoretical resolution is ~70 m though smaller objects of high contrast can be distinguished. The pixel (*pic*ture *el*ement) size is 79 × 56 m. With low contrast, objects of this size could not necessarily be distinguished. This does limit the utility of the scanners to the wider types of estuary. Following initial 'look-see' types of investigation in which ERTS pictures were compared with hydrographic and tidal charts (Klemas, Borchardt & Treasure, 1973; Klemas *et al.*, 1974) more sophisticated computer processing is being undertaken (Armstrong & Brimblecombe, 1975, 8).

Little has been done in the way of calibrating these sensors by 'sea-truth' measurements. This is likely to be a very difficult task requiring synoptic sampling from many ships. From general knowledge of an area reasonable values can be estimated (Klemas *et al.*, 1973) and it would be unwise to hope for much better for some time to come.

Nimbus-G

This satellite will carry a coastal zone colour scanner (CZCS). This will have five narrow bands in the visible part of the spectrum, each ± 0.01 µm, located at wave lengths of 0.443, 0.520, 0.550, 0.670 and 0.750 µm. There is strong absorption by green and red chlorophylls at 0.443 and 0.670 µm. One would expect use being made of the 0.670 µm and possibly the 0.550 µm bands for suspended sediment monitoring. The spatial resolution is rather poor, 800 m, making this sensor of limited use in many estuaries.

With both the LANDSAT and Nimbus-G data it must be borne in mind that it relates only to the upper part of the water column and that estuarine areas are generally stratified. It cannot be used alone to establish sediment transport budgets. Synoptic surface patterns, however, may be shown better by this than by any other method.

Determinations of properties

Particle size

Before examining the various methods available for determination of sediment size one must ask oneself *why* one wants to know the size and *what* size of the particles one wants to know. The usual contexts in which size distributions are sought are, (i) studies of sediment variations in relation to tidal dynamics and sediment budgets, (ii) studies of particle dynamics – aggregation, disaggregation, flocculation etc., (iii) sediment tracing and provenance, (iv) determination of suspended organisms – changes in their numbers and population-size characteristics. For any work on suspended sediment dynamics the most useful parameter is the *in situ* settling velocity distribution but this is rarely measured. Work on particle dynamics and organisms would probably be best served by methods where you could see the objects whose size was being determined. Tracing requires a method where very detailed distribution of the basic population of disaggregated particles can be made.

The methods available to us involve sedimentation, with gravimetric or optical measurement, electronic counting and sensing of a parameter roughly proportional to particle volume, sieving and measurement under the microscope. Not all methods of size analysis are available. Pipette analysis, the classical method for fine sediments needs typically 15 g l^{-1} and at least one litre of suspension at that concentration. This is not feasible in most cases. Even if it were possible to concentrate 15 g into one litre, it would not be an acceptable method of analysis. If settling velocity data are desired from sedimentation analysis then the suspension should be disturbed as little as possible from its condition in the estuary, i.e. no concentration, no shaking, or decanting. This requirement rules out most sedimentation methods but it leaves optical sensing which can deal with suspensions of low concentration. This section is concerned therefore with the analysis of *small amounts* of sediment or *low concentration* suspensions. The particular problems of sediment settling velocity analysis will be dealt with in the next section. Here we deal with microscopic measurement, electronic sensing and sieving. Many of these methods are familiar and have been usefully reviewed by Swift, Schubel & Sheldon (1972). There is not the space here to go into analytical details; simply the main features and uses of the methods will be given here.

Microscopic measurement and counting

This method has been used mainly by Schubel (1969) in Chesapeake Bay. He used a Zeiss Particle Size Analyser TGZ-3 to size images of particles displayed on photomicrographs. In addition Schubel & Schiemer (1967) built their own semi-automatic microscopic size analyser. Small samples of water are taken and filtered on cellulose ester membranes. These are clarified with an oil of the correct refractive index, mounted on a slide and photographed under the microscope. The Zeiss analyser has an iris diaphragm whose aperture size is related via a commutator to 48 telephone counters, each counter corresponding to a given aperture interval. Light shines through the diaphragm and a lens on to the photomicrograph. The light spot is matched in diameter to the average diameter of a particle on the photograph, a foot switch is depressed and another particle is then measured. An experienced operator can do 1000 particles per half-hour. Schubel made three slides from each sample and counted 1500 or 2000 particles. It was found that consistent values of the first two statistical moments of the size distribution were given by samples of 1000.

This device yields number frequency data. They may be converted to volume frequency data if the diameters measured are assumed to be those of spherical particles (not such a poor assumption considering the equally gross assumptions made in other methods of analysis) (Fig. 6.11). However in this case 1000 counts do not generate stable statistics (means vary by a factor of two) (Schubel, 1968, Table 9). Conversion to weight frequency data would require the additional dubious assumption of a constant value for particle density.

The advantages of this method are that with care the size distributions are virtually undisturbed from their state *in situ* (particularly if the sampling technique of Schubel & Schiemer (1972) is used) and that one can see the particles being sized to make some estimate of their nature. However it is important that the effects of particle settling in the bottle be nullified by gentle

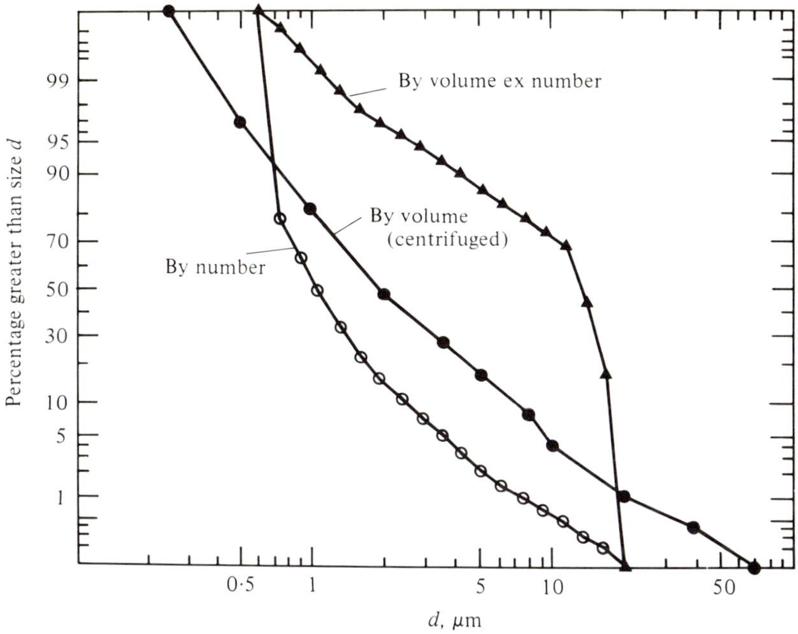

Fig. 6.11. Analyses of the same sample by counting particles (curves of number and volume ex number) and by centrifuge sedimentation analysis. It will be noted that conversion of counts of particles to volume does not give results similar to measurement of volume size distribution. The number values are effectively truncated at the high sizes because the effort of counting tens of thousands of particles would be too great. (From Schubel, 1968, Station IIIc, 22/8/66, mid-depth.)

mixing of the water; vigorous shaking may cause either flocculation and disaggregation of particles. The disadvantages are the time-consuming nature of the analysis, both in the production of photomicrographs and in the particle counting, and the fact that when converted to volume (a basis at least broadly comparable with other methods) the statistics for reasonable numbers of counts are not very stable. While it might be supposed that this method would be most useful in studies of particles and their properties, nevertheless Schubel (1969) has managed to show the relationships between physical processes and size distributions measured in this way.

Emphasis has been placed here on the systems which have been used. There are on the market other devices such as the Quantimet scanning image analyser which are automated and more rapid. However they have not been used for suspensions, though there is considerable potential for them.

Electronic particle sizing and counting

There are now a number of makes of electrical sensing-zone types of particle counter of which those made by Coulter Electronics and Particle Data Systems are the best known. The method, originally patented by Coulter, is that a constant current is applied across an aperture through which particles suspended in an electrolyte are sucked. This produces pulses in the voltage which are related to the particle volumes. The most recent version of the Coulter Counter records both the number and magnitude of the pulses, converts them and sums them as particle volumes and can deal with up to 5000 particles per second. The nominal volume units of the machine and the boundaries between the channels in which the data are accumulated, are related to actual particle volume and diameter via a series of calibration experiments, using spherical particles. Thus the output of the machine is of volume–size distribution. The channel sequence is based on 3 channels per doubling of diameter, the arrangement yielding a $\frac{1}{3}\varphi$ scale (Sheldon, 1969). The relationship of the measured distributions to particle diameter depends on the particles being effectively non-conducting and equant in shape. The first requirement may be a problem for loose aggregates containing perhaps $> 90\%$ salt water. The second requirement is a problem in the case of rod and thin-plate shaped particles. Little is known about the first problem, but experiments with particles of different shapes suggest that the latter is less acute than theory would suggest (Allen, 1968).

Details of the techniques for operation of the Coulter Counter are given in Sheldon & Parsons (1967) and a method for dealing with sand is given by McCave & Jarvis (1973). More recently a method has been devised by McCave (in press) for generating size distributions with points at $\frac{1}{9}\varphi$ intervals. This confers the advantage that the detailed modal structure of the size distribution is shown and this may be of use in tracing suspended material (Fig. 6.12).

The advantages of the method are many. It is rapid, reproducible and can be used on suspensions of low concentration (< 1 mg l^{-1}). More concentrated suspensions need to be diluted. It sizes the particles in their own salt-water medium and has a range from 0.5 to 850 μm. Disadvantages are that the equipment is expensive and very sensitive electronically. The latter factor plus the fact that it requires mains voltage means that it cannot be used on the small craft often used for estuarine work. These tend to be extremely noisy electronically. However, it is the experience of myself and my students and of Kranck (1975, p. 115) that suspended material size distributions are far more resistant to settling and resuspension by agitation than one would suppose. Material analysed after settling out and resuspension by gentle shaking or stirring gives the same results as analyses made on the original

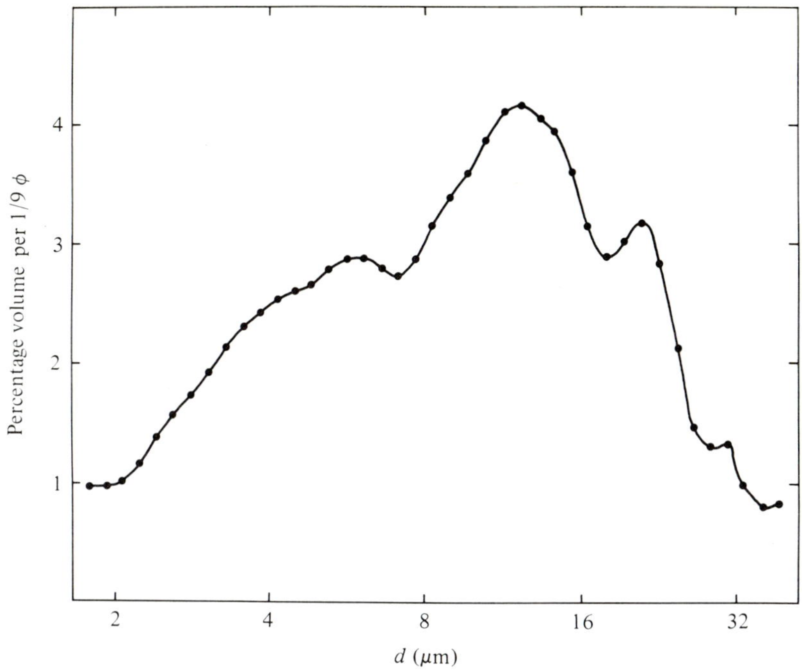

Fig. 6.12. Size analysis at 1/9 φ interval of suspended material from Yaquina Bay, Oregon, using a model T Coulter Counter. This is, in effect, a volume–frequency curve. The make up of the sediment in terms of populations with distinct modes is emphasised.

samples. It takes at least a minute in an ultrasonic bath to start to break down faecal pellets. So data of value may be obtained from samples analysed some hours after collection.

Sieving

Until recently this would not have appeared in a discussion of most suspended material work. However there is now available from Fritsch a set of micro sieves with aperture sizes at 5 μm intervals between 5 μm and 50 μm. They may be used in a wet sieving system on an electromagnetic sieve shaker. The system is not cheap, has not yet been used for suspended material but might be of interest. The larger mesh sieves in this range (or any standard set of 10-cm diameter sieves) can be used for analysis of small samples of sand taken from suspension. Samples as small as two grams could be analysed with acceptable accuracy (de Vries, 1970).

Sedimentation techniques

With some sedimentation methods the intention is to determine a particle-size distribution via the settling velocity, whereas with others an approach to the *in situ* settling velocity of distribution is sought. Two particular techniques for small amounts of fine sediment fall largely into the size-distribution objective.

The sedimentation balance is an old idea which has now been linked with a microbalance by Cahn and Sartorious. These systems analyse the settling velocity distribution of dispersed, disaggregated samples containing about 80 mg as a minimum, but with a concentration of more than 1600 mg l^{-1}. The machine gives a continuous chart record of the accumulating weight on the balance pan from the settling suspension. This is converted into a record of settling velocity by digitising the record at appropriate intervals. The settling velocity is converted to size by use of Stokes' Law. Very detailed instructions for operating a Cahn system are given by Thiede *et al.* (1976). The main advantage of this system is that it yields a continuous record. However it is expensive, sensitive to vibration and temperature change, slow (one sample per day) and requires a lot of sample preparation. However there are few methods of getting sedimentation diameter data for small samples.

Centrifugal settling is a proven technique. Schubel (1969, 1971*a*) used the MSA particle size analyser. This employs centrifuges with different rates of rotation and centrifuge tubes with capillary collectors. The diameters of these are 1 mm, 0.75 mm and 0.5 mm. The quantities of sample used are very small, a few milligrams. The suspension is made up in a tube and an initial period of gravity settling is allowed. The tube is mounted in a projector which magnifies the capillary and the rate of accumulation of material is measured. Then the tube is transferred to a centrifuge, spun for a preset time at known speed and another measurement of height is made. This is repeated using higher speed centrifuges for the finer fractions. The settling times are calculated using eqn (6.1) with, in this instance, the sphere of diameter d settling down a tube from r_1 to r_2. Details of the operation and calculations are given in Schubel (1968). The samples used by Schubel were filtered on to cellulose or silver membrane filters. Material was removed by brushing or solution in acetone (cellulose) and brushing or ultrasonic treatment (silver). It is claimed in a number of cases, on the basis of comparison with particle count measurements, that the naturally occurring size distribution was recovered and analysed (see Fig. 6.11). However where there was appreciable organic matter present peroxide treatment and complete dispersion was effected.

This method is not good for reliably estimating settling velocity distributions, and thus for standardisation the samples should be well dispersed ultrasonically with peroxide and dispersant treatments. Its great advantage is the very small amount of material required. However the Coulter Counter requires no more and is considerably faster, and measures the grains in their undispersed state. It should be noted that Kranck (1973, 1975) found that the log(grain size modal diameter) for dispersed material and log(floc size modal diameter) for the same material after flocculation were linearly related. So results from dispersed samples may still reflect some aspects of the hydrodynamic control of their parent aggregated population: bigger grains give bigger flocs!

Settling velocity distribution

The measurement of settling velocity distribution is intrinsically simpler than size distribution because there must be no sample pre-treatment and, indeed, as little interference with the sample as possible. The main differences between the methods are in the sensing of the suspended material settling rates.

Postma's 'pipette' method

In a study of suspended material in the Dutch Wadden Sea, Postma (1961*b*) took six-litre samples and poured them into an 85 cm tall cylinder. This he treated in the same way as in the standard pipette method for fine sediment analysis (see e.g. Krumbein & Pettijohn, 1938; Galehouse, 1971; Folk, 1974). First a sample was taken immediately after filling the cylinder, to estimate total concentration, then samples were siphoned at different depths and times to give weights of the fractions with settling velocities less than 0.38, 0.095, 0.023 and 0.006 cm s^{-1}. The 0.5 l samples were filtered to determine the weight of material in the fractions. The volumes taken by Postma were too large for accurate determination as nearly half the volume of suspension was taken and each sample of 0.5 l must have contained fluid drawn from well above and below the nominal sampling depth in the tube. Smaller samples of 100 ml containing 1 mg of sediment could be accurately determined using Nuclepore filters. Thus settling velocity distributions could be obtained from 5 l of suspension with $C \sim$ 10–20 mg l^{-1}.

Owen's bottom withdrawal tube

For work in the Thames Estuary Owen (1971) designed a tube (length 1 m, diameter 5.1 cm, volume 2 l) which samples the suspension in a horizontal attitude pointing into the current with the help of a tail fin. It then rotates into an upright position on breaking the surface. From this time settling is deemed to be occurring and the tube is used as a conventional bottom

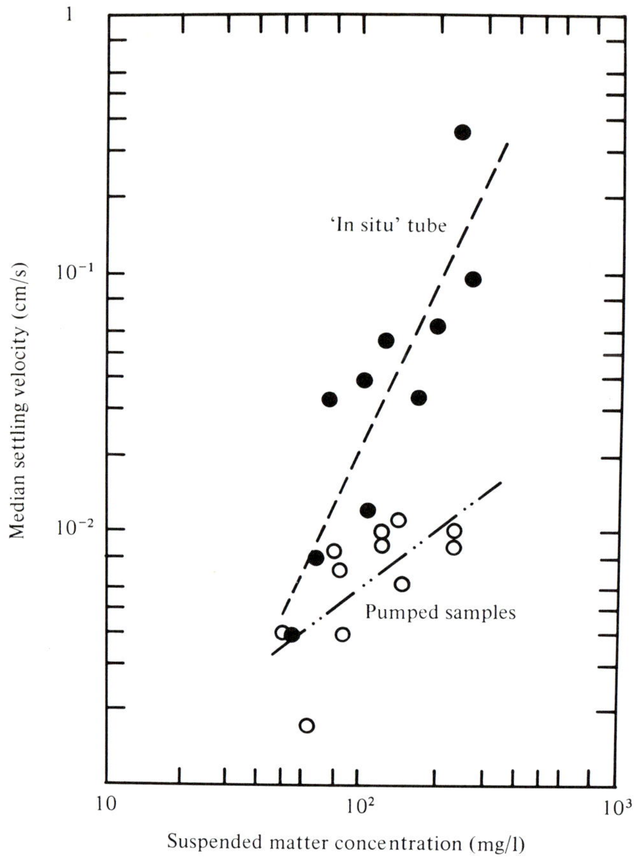

Fig. 6.13. Settling velocity of suspended sediment in the Thames measured by the Owen tube and by analysis of pumped samples. It appears that the larger particles (higher w_s) are broken up in the pumping procedure. (After Owen, 1971.)

withdrawal tube (Vanoni, 1969). This method is analogous to the pipette method except that the samples are taken from the bottom of the tube. The method has a similar potential for dealing with low concentrations of suspended material if used with precision gravimetric techniques. It is noteworthy that the results from this method are different from those given by samples obtained by pumping (Fig. 6.13). In this respect the decantation involved in Postma's method is probably best avoided by using the same tube for sampling and analysis.

Optical methods

Swift & Pirie (1970) made a number of shipboard grain-size measurements in the Gulf of San Miguel using a photo-extinction turbidometer. As shown above, the attentuation or scattering of a light beam is proportional to the concentration (among other parameters) of suspended material. The method employs this principle to sense the varying concentration of sedimenting material in a column. This may be done either at one height or at a variety of heights. Details of the theory, apparatus and calculations are given by Simmons (1959) and McKenzie (1963). Inaccuracies are present in the method because of non-sphericity and lack of uniform refractive index of the particles. However comparison with measurements on cleaned and dispersed suspensions yields good results (Simmons, 1959). The results shown by McKenzie and by Swift & Pirie only go down to 5 μm and most of the San Miguel samples have $\sim 50\% < 5$ μm. One cannot expect reasonable statistics from half the distribution. The minimum analysable concentration depends partly on the length of the light path through the suspension. Swift & Pirie (1970) report a minimum value of 70 mg/l. It is pointed out by Swift, Schubel & Sheldon (1972) that scattering is more sensitive at low concentration than attenuation. However no modern attempts have been made employing this principle. Apparatus is commercially available for laboratory work, adaptable for field use, from Microscal Ltd. The Microscal instrument has a lower size of about 2 μm, which improves the working range.

The method may be quite useful in more turbid estuaries but the concentration requirement may be a limitation in many. The main problem is that one can do only one sample at a time. This gives a practical limitation to no more than about one sample per hour. A number of Postma or Owen tubes can be made and operated on board simultaneously, but with more expensive equipment the opportunities for multiple sets are limited. As with the Coulter Counter most versions will need mains voltage, though a 12 or 24 volt version with a chart recorder could be constructed.

In situ *estimation of effective settling velocity*

The suspended load equation has been used by Krone (1972) to obtain an effective settling velocity. The equation describes the variation in concentration C of suspended material with height above the bed z in a flow of depth D. Measurements are related to a value of C at a height a just above the bed,

$$\frac{C}{C_a} = \left[\frac{a}{D-a} \cdot \frac{D-z}{z}\right]^{\zeta}. \tag{6.7}$$

The exponent $\zeta = w_s/\kappa u_*$ is the slope of the line when C/C_a is plotted against $(D-z)/z$ on logarithmic axes (Fig. 6.14). In this w_s is the settling velocity of

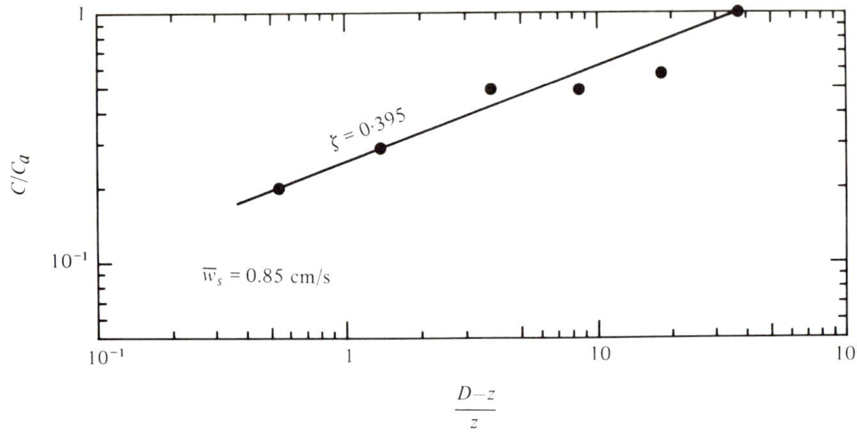

Fig. 6.14. Plot of the parameters of the suspended load equation yielding the exponent $\zeta = w_s/\kappa u_*$ as the slope of the line.

the material, κ is von Karman's constant (0.4 for clear water) and u_* is the shear velocity. The data are plotted as on Fig. 6.14. From the slopes the values of ζ are obtained. Simultaneously the velocity profile near the bed (lower 2 m) is measured. If it is logarithmic, it is fitted by the relation

$$\frac{u_z}{u_*} = \frac{1}{\kappa} \ln \frac{(z+z_o)}{z_o}, \tag{6.8}$$

where z_o, the bed roughness length, is the intercept on the $\ln z$ axis, and u_*/κ is the slope of the plot u_z against $\ln z$. The method is approximate and the value obtained relates to some concentration-weighted mean w_s. Values obtained by Krone in the Savannah River Estuary have means ranging from 0.46 cm s^{-1} to 3.35 cm s^{-1}, values four orders of magnitude greater than those for un-flocculated clay. These values were obtained with $\kappa = 0.4$. Even if it is assumed $\kappa = 0.2$ as it might be in high concentrations of suspended sediment, these settling velocities will be diminished by a factor of four still leaving 0.11 to 0.84 cm s^{-1} as the floc settling velocities. Measurements of settling suspensions from the Chao Phya, another turbid estuary, yield values up to 100 times the settling velocity of entering river sediments (Allersma, Hoekstra & Bijker, 1967).

The Gry tube for sand

Sedimentation analysis of small quantities of sand is not often required or undertaken. However work with pumping systems or the Delft Bottle may yield rather small samples. The Gry tube (Gry, 1941) with modified introduction system (Postma, 1965) is a water-filled 1 m glass tube tapered and with a tap at the bottom. Sand is introduced at the top from a closed water-filled chamber by turning a tap, and settles down the tube. At pre-determined times evaporating dishes, containing a little water, are placed under the bottom end of the tube. The sand sinks into the dish and is removed. In this way the settling velocity fractions are removed for weighing and microscopic or other determinations. Determinations giving seven points can be made on samples of 100 mg or less. The method is tedious in that many evaporating dishes have to be accurately weighed, dried and weighed again. The alternatives are the Coulter Counter, which is not optimal for sand, or pumping a larger sample for sieving, neither of which give the settling velocity distribution.

Density of particles

Where particle volume data is acquired, as in the case of Coulter Counter analysis, one could make some estimate of the settling velocity distribution if the *in situ* particle bulk density was known. There are very few data bearing on this problem; most are reviewed in McCave (1975). Most of the data by Krone (1963, 1976) are mainly estuarine and comprise inorganic floccules. These range in density from 1.25 g cm^{-3} to 1.056 g cm^{-3} when saturated with salt water (ρ = 1.025 g cm^{-3}). When plotted as density contrast with seawater, it is seen that most determinations show decreasing density with increasing

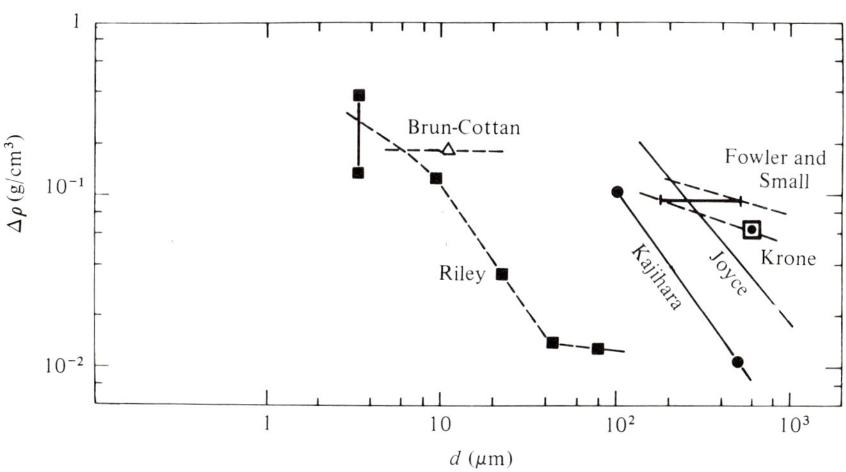

Fig. 6.15. Plot of the density difference $\Delta\rho = (\rho_s - \rho)$ between suspended sediment and water as a function of size, mainly from McCave (1975). Individual sources are Riley (1970), Brun-Cottan (1971), Kajihara (1971), Joyce (1976), Krone (1972), and Fowler & Small (1972), for pellets. Noteworthy is the fact that the slope of $\Delta\rho$ against d is similar for Riley's data on mainly organic aggregates, Kajihara's aggregates and Joyce's faecal pellets from *Pectinaria*.

size, but a considerable range of densities for a given size (Fig. 6.15). Most of the values are < 0.3, i.e. particle density is generally < 1.325 g cm^{-3}, far less than the primary particle density of mineral matter. Only Schubel (1968) records particle densities in excess of two, his values, measured with a pycnometer, being 2.24 to 2.6. In addition to inorganic flocs, organic/mineral aggregates are common. Both types have generally low densities. Apart from the pycnometer the most common measurement technique has been to measure settling velocity of quasispherical particles and obtain the density from Stokes' Law. Krone (1963) obtained his values from viscometer measurements. However the basic data on density distribution with size are still lacking for a variety of estuarine suspensions.

Organic matter

The determination of the 'organic matter' content of suspended material is a basic operation carried out in most studies. The principal method used is combustion. However, modern equipment has changed the more precise methods of determination of particulate organic carbon (POC) and nitrogen into a routine operation. This section outlines the main methods in use.

Ignition loss

The combustion method is described by Manheim, Meade & Bond (1970). Sediment was collected on cellulose ester membranes. These have a negligible ash weight. The filter is cut in half, pre-charred with alcohol in a platinum crucible, ashed at 500 °C and weighed. The original and post-filtration weights of the filters and blanks are known, thus the combustion weight loss can be calculated. This is some measure of organic matter. Many people have presented factors which indicate that combustion loss is about double the organic carbon content. Meade *et al.* (1975) determined combustion loss while a parallel series of samples was run to estimate particulate organic carbon. The ratio was combustion loss = 1.6 POC, with a large scatter. The difference is probably due to the ignition driving off some of the bound water from the clay minerals.

Peroxide treatment on Nuclepore filters

This method uses the fact that there is near-perfect removal of suspended material from Nuclepores with ultrasonic cleaning. Samples are taken, filtered and weighed as usual. The filters are then transferred to clean, small, glass bottles and covered by 35% hydrogen peroxide buffered by a few drops of ammonium hydroxide. They are then placed in a low power ultrasonic bath (50 watts at 20 kHz) for 30 s. The filters are removed, rinsed over the mouth of the bottle with filtered distilled water and returned to their individual petri dishes. The bottles have a screw cap loosely put on (or a foil cover) and are then put in a 60 °C oven overnight. The peroxide with sediment is then refiltered *through the same filter from which it came*. The filters are dried, weighed and the weight loss (peroxide-degradable fraction) determined. Blanks for the procedure show weight changes within the limits of error for the difference in two weighings of a filter. The ultrasonic treatment probably aids the oxidation by breaking down the organic matter. It may well also break down some of the mineral matter to less than 0.4 μm allowing it to pass the filter. Consequently the values obtained are likely to be over-estimates. However, this and combustion are both rapid approximate methods. The end-product of the peroxide method is a filter loaded with a cleaned, disaggregated sediment which may then be analysed by Coulter Counter for basic grain-size characteristics (see Fig. 6.12).

CHN analyser

The CHN (carbon, hydrogen, nitrogen) analyser is made by Hewlett Packard and Perkin Elmer. The most common method of analysis for carbon has been that of Menzel & Vaccaro (1964) in which the samples are combusted on glass fibre filters and the evolved carbon dioxide analysed by an infrared gas analyser. Current CHN analysers are similar in that the sample is filtered on precombusted glass-fibre filters and subjected to controlled pyrolysis with analysis of the gas on a chromatograph. Details of methods for dealing with the small quantities encountered in suspended material analysis are given by Sharp (1974).

Quantitative wet oxidation techniques

Strickland & Parsons (1968) give many methods for dealing with various aspects of particulate materials. Material filtered on to a glass-fibre filter is oxidised wet by a mixture of potassium dichromate and sulphuric acid. This bleaches the dichromate and the amount of bleaching is measured in a spectrophotometer with reference to the pure oxidant. The method is fairly rapid and could be carried out at sea. The carbon determined is as glucose carbon.

Composition

Most of the techniques of analysis mentioned in this section are described in detail in standard reference works. Here it is simply intended to draw attention to the principal methods and any special equipment or techniques available for dealing with suspended material.

Visual estimation by microscopy

Light microscope. A number of studies of suspended material have used the technique of clarifying a cellulose ester membrane and examining the material using a petrographic microscope (e.g. Manheim *et al.*, 1970; Conomos & Gross, 1972). The techniques are the normal ones of biological and mineralogical microscopy for which standard handbooks are available.

Scanning electron microscope. There are enormous problems of statistical variability in the analysis of small numbers of particles in small fields of view at very high magnification. The preparation of the stubs for viewing is time consuming, so that great time and effort would be involved in gaining reliable results. In consequence most studies have simply included a few SEM photographs to demonstrate the presence of particular components (e.g. Schubel, 1971*b*). However there is certainly scope for more extensive work both in identification of particles and especially in examination of particle aggregation. In this respect the techniques of Harris *et al.* (1972) and Schubel & Schiemer (1972) are available for mounting and sampling and freeze-drying for SEM analysis.

X-ray fluorescence analysis

Where only small quantities of material are available, thin-film XRF methods are appropriate. A method was first outlined by Cann & Winter (1971) for estuarine sediments and used in an extensive set of analyses by Frostick (1975) in the estuary of the River Deben in Suffolk. Recently, variants have been described for use with deep-sea samples by Baker & Piper (1976) and for estuarine samples by Subramian & d'Anglejan (1976). The principal use for this is for major elements, aluminium, silicon, calcium, potassium, titanium and iron.

X-ray emission spectrometry has been used in a number of estuarine studies for both major and trace elements (Price & Calvert, 1973; Wollast *et al.*, 1973; Price & Skei, 1975). Devices for measuring emission spectra can be incorporated in scanning electron microscopes allowing analysis of individual particles as seen under SEM.

Neutron activation analysis

The powerful technique of activation analysis may be used for suspended material. The filters with suspended material on them are encapsulated in plastic irradiation vials and neutron activated together with standards of the elements of interest. Techniques are described in Bertine *et al.* (1972), and general principles are in Kruger (1971) and other texts.

Leaching experiments

Attempts to find where certain elements are attached in suspended material, and how strongly, generally involve leaching experiments and re-analysis. Leaching has been carried out with EDTA, hydrochloric acid, acetic acid, and other treatments, as discussed by Chester & Hughes (1967). Either the sediment is re-analysed or the leachate is analysed to determine the components removed.

X-ray mineralogy

There are two important features about X-ray diffraction work on suspended sediment: it is often performed directly on the original filter, and normally very small quantities of sediment are involved. When material is analysed as filtered it contains sand grains, inorganic matter and other debris and the diffractograms are almost invariably of poor quality (e.g. Manheim *et al.*, 1972; d'Anglejan, 1970). However it is rapid and useful for establishing presence or absence of specific minerals. Silver filters have been most commonly used (for technique see Manheim *et al.*, 1972) but cellulose ester membranes can also be used. These give a broad bulge on the diffractogram with mineral peaks superimposed. Nuclepore filters give an even sharper bulge and do not form suitable substrates for diffraction analysis. For more accurate clay mineral work it is necessary to clean the sample of organic matter with buffered peroxide, separate the 2 μm fraction and prepare mounts in an appropriate way. If the amount of sample on the mount is < 1.6 mg cm^{-2}, Tucholke (1974) reports that montmorillonite will be underestimated. In order to achieve the correct loading and separation of clay size it will probably be necessary to have a gram of material available, so that in some estuaries harvesting will be necessary.

Interpretation of results

In this section I want to give some of the objectives of estuarine investigations to which the techniques described contribute. This will be done by examining particular aspects of estuarine behaviour and seeing how the data are used and have been used in the more noteworthy studies. The latter are those which are known and published. Because pollution of estuaries is a matter of public concern much work is carried out around the world in government and industrial laboratories. Only a fraction of this material is published in the open literature, the rest remains in internal reports with limited distribution. Nevertheless the most wide-ranging studies of estuaries have been those undertaken by government laboratories and consulting firms. They have the commitment and the finance to mount studies lasting for years and requiring the deployment of a number of survey vessels.

Concentration profiles

The distinction is made in fluvial work between 'wash load' and bed-material load. The former is of a grain size appearing in only very small quantities in the bed of the river but which comprises fine material almost permanently in suspension. Bed-material load is the general material of the bed which is transported along the bed, and in suspension when the flow is strong enough. In practice this distinction boils down to silt/clay or mud on one hand and sand on the other. Because of its smaller settling velocity the mud is more uniformly distributed through the water column while the sand is concentrated near the bed.

There are two principal objectives in obtaining concentration profiles: first, they are necessary for computation of fluxes; the product of concentration and water speed over the whole flow depth. The other use is to determine some property of the suspended material (as in the example given on p. 158 for calculation of settling velocity for calculation of settling velocity) or, if those properties are known, some dynamic property of the fluid, such as the eddy viscosity.

Fine suspended sediment

An example of profiles of both coarse and fine suspended material is given in Fig. 6.16 taken from Thorn (1975*b*). The points may be fitted by a curve drawn by eye or, preferably, using the suspended load equation (eqn 6.7).

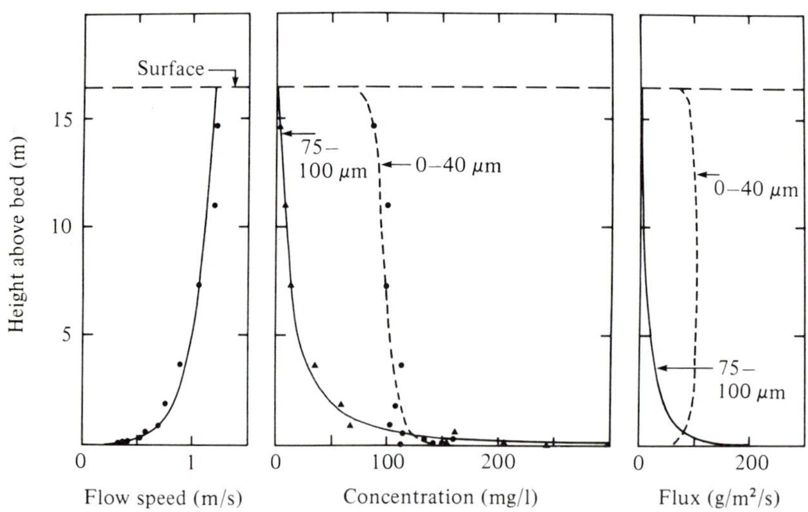

Fig. 6.16. Profiles of flow speed, concentration for two size ranges, and flux for these two ranges for a station in the Outer Thames Estuary. (From Thorn, 1975*b*.)

However, w_s and u_* in the parameter $\zeta = w_s/\kappa u_*$ are generally not known. In some cases u_* may be approximated from a value of current speed and an appropriate drag coefficient (see Sternberg, 1968). w_s may be known or obtained by Krone's method from the best fit line.

The distribution of suspended material has been used by Joseph (1954) to calculate the eddy diffusivity for sediment. For stationary conditions and values of $\zeta < 1$ an approximate value for the eddy viscosity of the fluid may be obtained from concentration and settling velocity data. However sediment is not good for this as it takes an appreciable time to diffuse upwards or downwards, and as the tidal currents in an estuary are either accelerating or decelerating continuously the condition of stationarity is not achieved. Joseph (1954) gives an example for fine sediment of $w_s = 0.07$ cm^{-1} ($d = 30$ μm quartz density sphere) for which the upward diffusion velocity of a turbid cloud would be 0.2 cm s^{-1}. In water of 10 m depth the cloud takes an appreciable fraction of the tidal cycle, 80 min, to reach the surface. This is an estimate of the time required for the fine-grained suspended sediment distribution to come to equilibrium with the tidal current speed. However, in that time the velocity has changed and instantaneous equilibrium is not attained. This leads to a time lag between the peak current speed and the peak concentration which might be interpreted as a lag in the diffusivity. This is not the case; it is the non-stationarity of the system which is the cause. Krone's (1972) calculation of settling velocity (p. 30, Fig. 6.14) is suspect for the same reason. Concentration profiles were obtained from a tidal estuary and interpreted using a theory (eqn 6.7) in whose derivation stationary conditions are assumed. The results may be reasonable approximations, no more.

Suspended sand

In most cases sand ($w_s > 0.4$ cm s^{-1}) is concentrated close to the bed and most suspended sand movement occurs in the lowermost metre or two of the flow. Analysis of material by the Gry tube yields good values for the settling velocity distribution. This has been used by Postma (1965) to find the vertical distribution of concentration of a narrow size range ($2.7-2.8\varphi$, $w_s = 1.54$ cm s^{-1}). From the vertical distribution of concentration he used Joseph's method to find the eddy diffusivity as $A_z = w_s/d(\ln C)/dz$. Even in this case the diffusion of a cloud to the surface in 10 m of water takes 11 min so some apparent lag effects may be found (though, it may be pointed out, some of the lag effect is probably real as turbulent intensities are greater for a given velocity on the decelerating tide than on the accelerating tide (Gordon, 1975)). Postma's values of $A_z \sim 100$–200 cm^2 s^{-1} for maximum current speeds of 60–85 cm s^{-1} are quite reasonable.

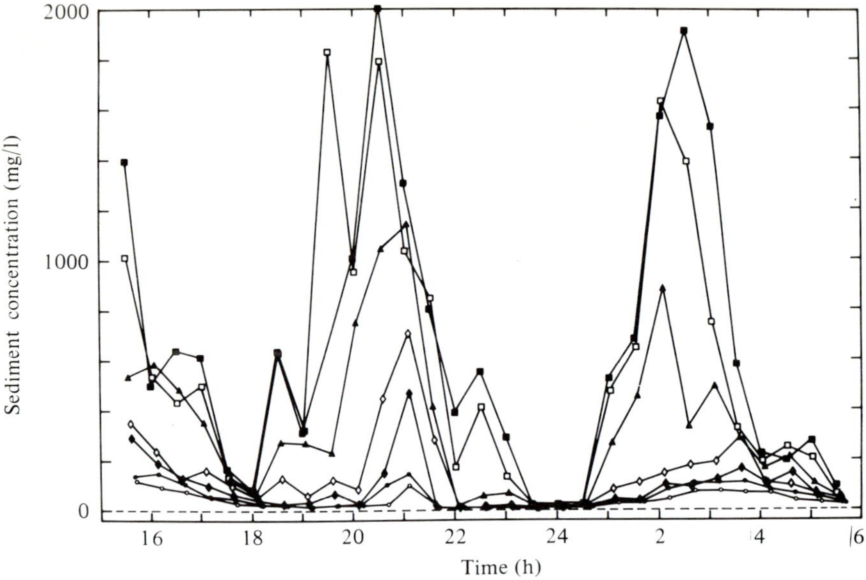

Fig. 6.17. Temporal variation of concentration at a variety of levels above the bed at a station in the Savannah River (from Krone, 1972). ■, 1 ft above bottom; □, 2 ft; ▲, 4 ft; ◇, 8 ft; ◆, 16 ft; ●, 25 ft; ○, surface.

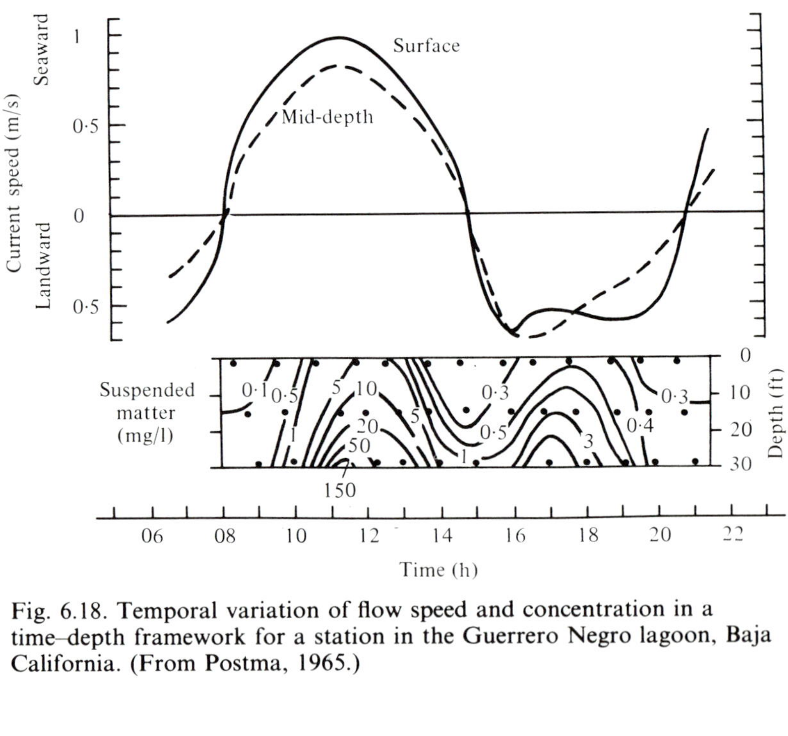

Fig. 6.18. Temporal variation of flow speed and concentration in a time–depth framework for a station in the Guerrero Negro lagoon, Baja California. (From Postma, 1965.)

Plotting of data

Concentration profiles are rarely obtained singly as in Fig. 6.16; they are normally part of a series taken at one station or a number of stations. Data from one station may be plotted as a succession of curves, such as Fig. 6.16, spaced along a horizontal time axis. More commonly they are plotted on a concentration–time diagram with the depths of different measurements denoted (Fig. 6.17). Alternatively data may be plotted on a depth–time diagram and contoured for different concentrations (Fig. 6.18). On this figure note that each point was sampled at a different time, therefore concentration profiles were not determined. They may be estimated from such a diagram though, by dropping a vertical at the desired point on the T-axis and reading concentration against depth pairs of values from the contours.

Fluxes

Total solids

As shown in Fig. 6.16 the flux for a section in terms of weight per unit width/time is given by the depth-integrated product of current speed and suspended material concentration. This simply gives an instantaneous value which is of little use. Flux calculations need to be made over tidal cycles and related to tidal range (Fig. 6.19) for analysis of sediment circulation or budgets. It is found that with increasing range there is increasing flux. It will be seen below that there are other controls on concentration (therefore flux) and the estimation of long-term budgets is complicated. The total flux per tide through a cross section is obtained by integrating the station value over a tide and multiplying by the estuary width by the station. This may lead to error in wide, vertically homogeneous estuaries where there is a horizontal circulation. For example Inglis & Allen (1957) show net flood transport on the north side of the Thames in Long Reach and net ebb transport on the south side.

The other principal problem associated with flux calculations relates to determination of net flux. Often the ebb flux and the flood flux are large numbers having a relatively large uncertainty. For example Postma (1961*b*) gives the transport through the Borndiep on the flood to be 66×10^5 kg. This is at least $\pm 5 \times 10^5$ kg. However in order to maintain the accretion of tidal flats only 3% of this ($\sim 2 \times 10^5$ kg per tide) is needed. Such a conclusion could never be arrived at from subtracting the ebb flux from the flood flux; the errors on the figures are just too great (McCave, in discussion of Meade, 1974).

The thorny problem of net fluxes has figured importantly in a number of studies including Inglis & Allen (1957) on the Thames, Allen & Castaing (1973) on the Gironde, Terwindt (1967) on the Rhine, Flemming (1970)

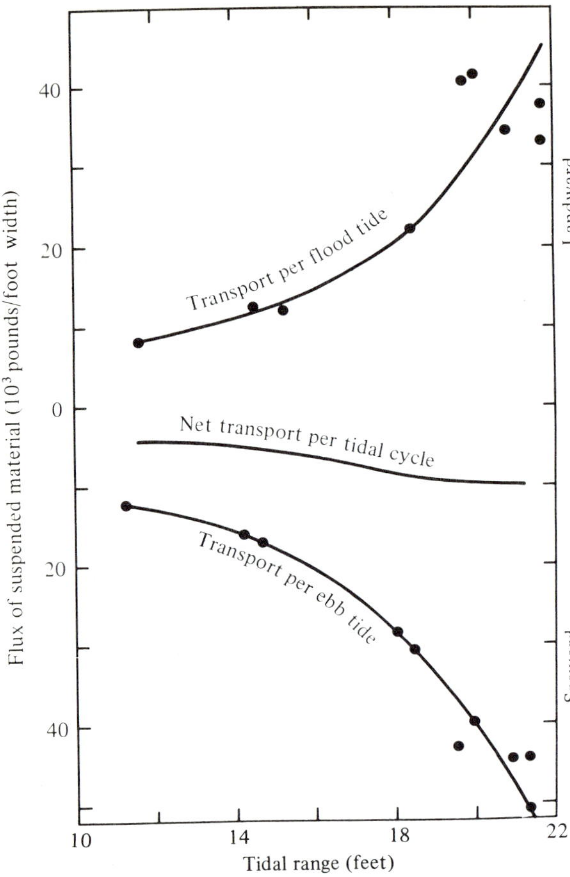

Fig. 6.19. Flux of suspended sediment as a function of tidal range for a station on the south side of Long Reach in the Thames. (From Inglis & Allen, 1957.)

on the Clyde and Cassie *et al.* (1962) on the Tyne. In the United States, Meade (1969) argues for net landward flux in a number of east-coast estuaries and fluxes are discussed for the Rappahannock by Nichols (1974). The Delft Hydraulics Laboratory has been active overseas with studies of the Demerara (1962) and the Chao Phya (Allersma, Hoekstra & Bijker, 1967).

Chemical and other components

It is of both scientific and social importance to know the fluxes of particular components in estuaries. It is no use knowing that there is net sediment escape from an estuary if one particular component carrying an adsorbed pollutant is being trapped in the estuary. In the calculation of budgets for chemical components it is necessary to have measurements of the concentrations in solution as well as in suspension. This is because some components may be adsorbed or desorbed from suspended material or degraded, as in the case of organic carbon. There are very few such budgets in existence. Wollast *et al.* (1973) have constructed one for the Schelde. From this it emerges that in a zone 30 km long above and below Antwerp 1.9×10^6 tonnes of material are deposited annually and of this 79×10^3 tonnes are organic carbon, 350 tonnes are lead, 1810 are zinc and 320 tonnes are zinc. Both solid and dissolved fluxes and pollutant input from the port of Antwerp are important items in the budget. Of 396 tonnes total input of copper only 63 tonnes escapes downstream, 50 in solution.

A further complication in the whole situation relates to the cycles of nutrients and plankton. There have been few budgets calculated for these though Postma (1954) made an early series of estimates for organic matter and phosphorus in the Wadden Sea. Conomos *et al.* (1972) and Conomos & Gross (1972) have examined various suspended material–plankton–nutrient aspects of the Columbia River Estuary but they have not presented a worked out budget. It is an extremely complicated task but nevertheless ultimately worthwhile. The estimation of fluxes and calculation of a budget forces us to make use of what may be meagre data and shows up their weaknesses. It also often shows us the way ahead.

Deposition

The loss of material to the bed by deposition represents another flux of suspended sediment in the system. Broadly this is controlled by sediment concentration, settling velocity and fluid shear. The full expression for the concentration of a flocculating suspension undergoing deposition according to Krone (1962, 1976) is

$$\ln\left(\frac{C}{C_o}\right) = -\frac{w_s t_c}{D}\left(\frac{n_p}{n_f}\right)\left(1-\frac{\tau_o}{\tau_c}\right)\ln\left(1+\frac{t}{t_c}\right), \tag{6.9}$$

where C_o and C are the initial concentration and concentration at time t, D is the mean depth of the depositing suspension, τ_o is the bed shear stress and τ_c is the critical shear stress below which all material deposits, n_p/n_f is the ratio of the number of primary particles to the number of flocs (i.e. the mean number of particles in a floc) and t_c is the 'coagulation time' the mean time between collisions of a particle. When $t \ll t_c$ (i.e. low rate of flocculation, low concentration) eqn 6.9 becomes

$$\ln\left(\frac{C}{C_o}\right) = -\frac{w_s t}{D}\left(1-\frac{\tau_o}{\tau_c}\right). \tag{6.10}$$

Differentiation of eqns (6.8) and (6.10) yields corresponding expressions for the rate of deposition in units of mass/area/time thus

$$R_f = -Cw_s\left(\frac{n_p}{n_f}\right)\left(\frac{1-\tau_o/\tau_c}{1+t/t_c}\right) \tag{6.11}$$

and

$$R = -Cw_s(1-\tau_o/\tau_c). \tag{6.12}$$

The first of these is difficult to apply as the time t from the beginning of flocculation cannot be simply expressed in a natural system. The term n_p/n_f can be eliminated by using Krone's (1962) approximate relation when $t \gg t_c$,

$$\frac{n_p}{n_f} \simeq \left(\frac{t}{t_c}\right)^2. \tag{6.13}$$

Krone (1962) argued that settling velocity should be proportional to the 4/3 power of concentration but Owen (1971) found $w_s = KC^n$ with $n = 1$ during spring tides and $n = 2$ during neap tides. So rather than use the longer form (eqn 6.11) Odd & Owen (1972) use

$$R = KC^{(n+1)}(1-\tau_o/\tau_c). \tag{6.14}$$

This is only an empirical approximation and measured values of w_s are to be preferred. Experiments show the parameter τ_c to be between 0.4 and 0.8 dynes cm^{-2}. These equations (6.12 and 6.14) appear to work satisfactorily for prediction of deposition in estuaries (Owen, 1977).

The dependence of rate of deposition on concentration is clear and this implies that the maximum rate of deposition should occur in the region of the turbidity maximum. That this in fact is the case is shown by studies of a number of estuaries (e.g. the Thames). Of course *net* deposition is the balance between deposition and erosion integrated over time. The mass rate of erosion dm/dt (mass $area^{-1}$ $time^{-1}$) is

$$\frac{dm}{dt} = M(\tau/\tau_e - 1), \tag{6.15}$$

where τ_e is the critical erosion shear stress and M is an erosion rate constant (Partheniades, 1965). The critical erosion stress has been measured by Migniot (1968) for many muds and he finds

$$\tau_e = 0.25\tau_y \quad \text{for} \quad \tau_y > 1.5Nm^{-2} \tag{6.16}$$

and

$$\tau_e = 0.32\tau_y^{\frac{1}{2}} \quad \text{for} \quad \tau_y < 1.5Nm^{-2}, \tag{6.17}$$

where τ_y is the yield strength of the material. Thus erosion is principally dependent on stress and the degree of consolidation of the mud. Use of these relationships will allow appraisal of the long-term balance of erosion or sedimentation along the estuary and thus whether the bed is a source or a sink in budgetary calculations.

Modelling

In order to avoid the necessity of continually having to go out and measure parameters of interest in estuaries, increasing attention is being paid to numerical modelling of estuarine behaviour using digital computers. In this a series of differential equations describing flow and behaviour of components of interest are solved for discrete time steps. The results of a simulation are compared with data and adjustments are made to constants or relationships used in the model. This powerful technique is now considered a necessary adjunct to any comprehensive estuarine investigation. Few models concerned with suspended materials have been devised. There is one for the Thames (Odd & Owen, 1972; Kendrick, 1972). General problems of modelling of mud behaviour are reviewed by Owen (1977). As mentioned previously, organic matter is an important component of estuarine suspended material. It is not only transported, but grows and decays. The calculation of fluxes in this case is complicated and enters the field of ecosystem modelling. Estuarine phytoplankton models are reviewed by Di Toro *et al.* (1977).

Spatial variation

As was mentioned at the beginning of this chapter, estuaries are more muddy than the waters upstream or downstream of them. The primary reason for this, as was demonstrated by Postma & Kalle (1955), is hydrodynamic trapping of suspended material. This is not the only aspect of spatial variation, other aspects are the distribution of different components and their use as tracers.

The turbidity maximum

The term turbidity maximum applies to the zone of highest mean concentration of suspended material whether measured optically or gravimetrically. Some other mechanisms have been suggested (Buller, Green & McManus, 1975) but the great majority of turbidity maxima as found in the Thames, the Elbe, Ems, Gironde, Rappahannock, York River, Savannah and San Joaquin/Sacramento system to name but a few, are caused by the net non-tidal circulation of the estuary and are located at the null zone. An example is given in Fig. 6.20. The null zone is the region with a mean near-bottom current speed of zero, upstream of which the residual flow is downstream, and vice versa. Thus the zone is a sediment trap.

The location of the null zone differs according to estuary type. In the salt-wedge type, dominated by advective processes, the turbidity maximum

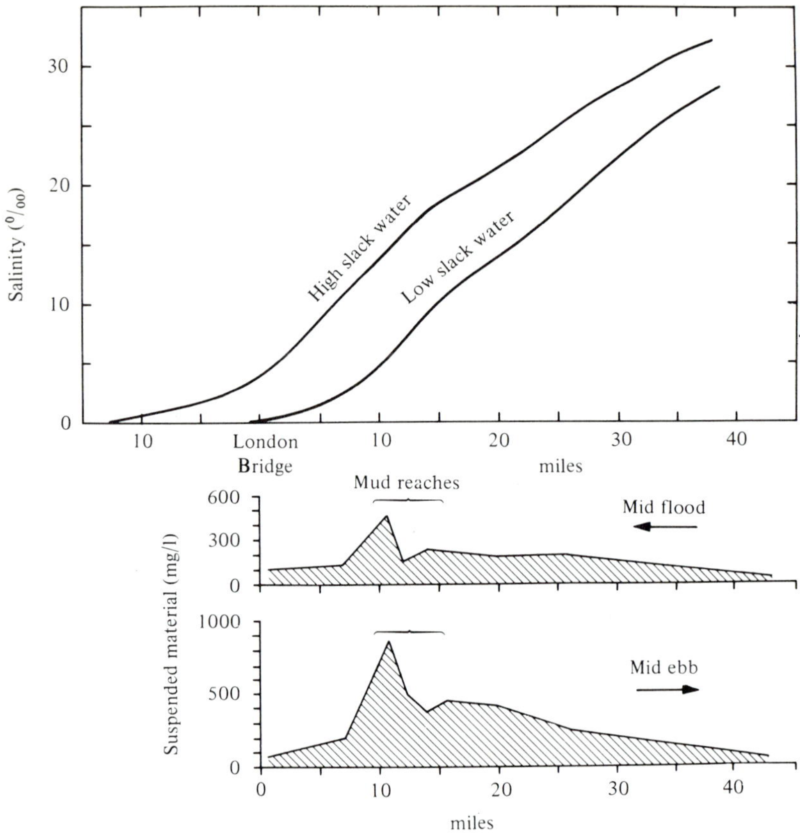

Fig. 6.20. Distribution of salinity and suspended material in the Thames, a well-mixed estuary for much of the time. (After Inglis & Allen, 1957.)

occurs at the upstream tip of the salt wedge at which the depth-mean salinity is very low. As diffusive processes become more important in partially mixed and vertically homogeneous estuaries so the null zone and the peak of the turbidity maximum occur at higher values of mean salinity (Fig. 6.20). In a sense both salt and suspended material diffuse upstream from the limits they would reach in a purely advective system. In fact, because of lag effects connected with deposition and resuspension, suspended material can 'diffuse' upstream into perfectly fresh water as long as there is still tidal motion. This is shown by Postma (1967) in an example from the Demerara River.

There are two noteworthy processes associated with the development of turbidity maxima. Postma (1967) suggested that they should be dominated by a narrow spectrum of grain sizes (settling velocities) because particles too fine to settle much would be swept out to sea in the surface plume, and large particles would sink and be deposited. Schubel (1969) examined this hypothesis in the light of data obtained in Chesapeake Bay and found it valid. Although a number of authors maintain that the mixing of fresh and saline waters will always lead to formation of flocs and that this will be an important process in the turbidity maximum, neither Schubel (1968, 1969) nor Nichols (1974) find this to be the case. However Schubel does find larger particles present but they are organic aggregates. Krone (1972) on the other hand argues that the high settling velocities of the material in the turbidity maximum of the Savannah River must be due to flocculation, but does not have pictures to prove it: a clear case for the use of scanning electron micrographs. Flocculation is related to concentration and to the shear stresses in the water. When the salinity is greater than about 2‰ the particles will cohere on collision and this is brought about by three mechanisms, all of which are concentration-dependent. (i) Brownian motion: This is most important for submicron particles. (ii) Fluid shear: at low shear rates collision is enhanced, but at high shear rates the flocs can be physically disrupted. (iii) Differential settling: the larger particles settling faster than the smaller will grow in size.

The dependence on concentration means flocculation should be most intense in the turbidity maximum, but near the bed the shear may be large so some flocs may be broken up in that zone.

In the turbidity maxima of a number of estuaries, 'fluid mud' has been detected. This material named by Inglis & Allen (1957, p. 843) in the Thames, but also known by the Dutch in Surinam as 'slingmud' and by the French as 'crême de vase', has a concentration in excess of $10\ g\ l^{-1}$ and generally around $100\ g\ l^{-1}$ or more. The relationships between the formation of fluid-mud pools and estuarine hydrography have been best demonstrated by Allen and co-workers in the Gironde (Allen, 1971; Allen, Castaing & Klingbiel, 1974). They are found to form in the null zone during neap tides. During spring tides the mud may be put back into suspension. The zone of fluid-mud formation is shifted downstream with the null zone during periods of high river discharge and upstream in low discharge. The fluid-mud pools may be a few metres thick, a hundred metres wide and kilometres long. Such an accumulation could contain from 10^4 to over 10^6 tonnes of mud and would be a dominant item in any sediment budget. Thus it is most important to search for possible fluid-mud accumulations using the echo sounder and samples. It almost goes without saying that fluid mud will play a dominant role in estuarial shoaling.

Component distribution

There has been little work on the distribution of components of suspended material. Conomos has distinguished biogenic components in the Columbia River (Conomos & Gross, 1972) and the San Joaquin–Sacramento system in San Francisco Bay (Conomos & Peterson, 1974). There is high organic

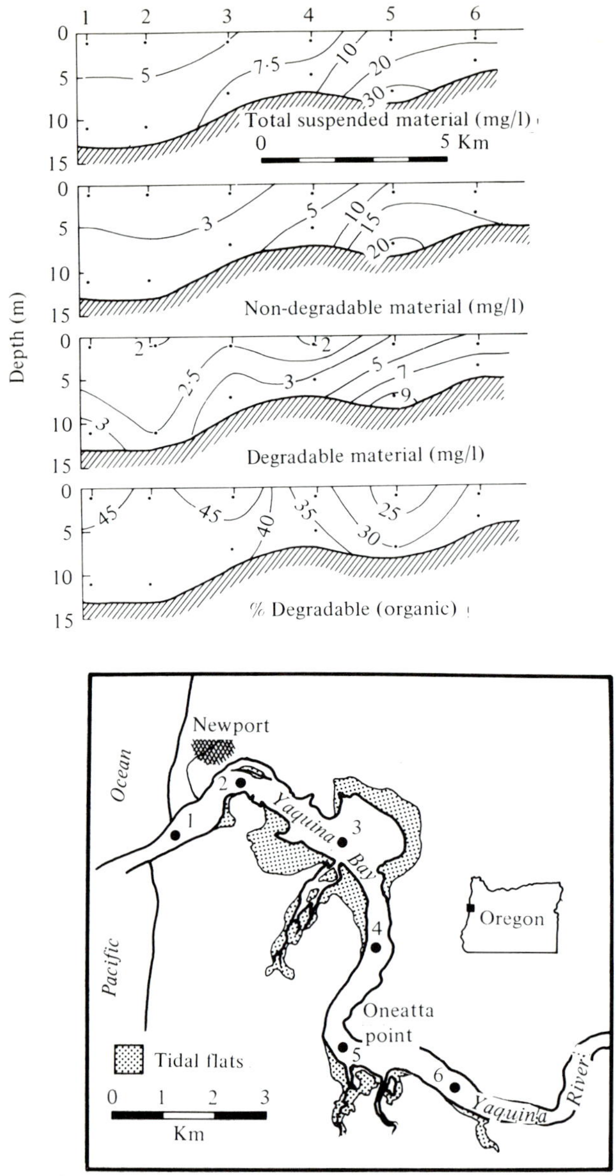

Fig. 6.21. Distribution of total suspended material, peroxide degradable ('organic') and non-degradable fractions in a single survey of Yaquina Bay, Oregon.

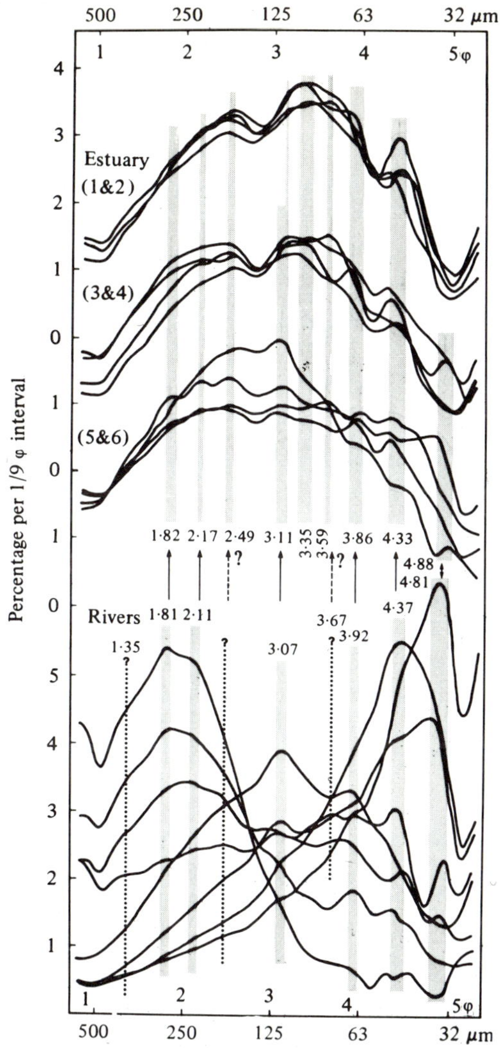

Fig. 6.22. Size analyses at 1/9 φ intervals (cf. Fig. 6.13) of suspensions from stations in Yaquina Bay (see Fig. 6.21. for locations) and the rivers supplying it. Overall the estuarine suspensions have most material in the medium-size ranges of 63 to 125 μm whereas the river suspensions show great variability. In detail the shaded bars pick out the location of modal peaks occurring in two or more samples. These match up well between river and estuary except the peak around 98 μm (3.35φ). This may indicate some supply from the sea as the peak is best developed in the more seaward stations, and is absent in the rivers.

concentration in the summer turbidity maximum thought to be caused by high plankton production in the null zone. Simple organic matter data can be rapidly obtained by combustion or peroxide techniques described above. Fig. 6.21 shows per cent organic matter in Yaquina Bay, Oregon obtained by the latter method.

Other bases for distinguishing components may be chemical or physical. Frostick (1975) examined the chemistry of fluvial and estuarine components in the Deben in Suffolk, but few other authors have made detailed examination of the spatial variation of chemical components. Detailed size analysis of the basic particle population also provides a way of distinguishing components. Size distributions of suspended material in Yaquina Bay and the rivers supplying it are shown in Fig. 6.22. Some components in the bay are not present in the river suggesting their introduction from the sea, a conclusion also reached for sands of the bed by Kulm & Byrne (1966). There is some scope for extension of these methods. They may provide a method of tracing the movement of components and assisting us in understanding estuarine behaviour.

Tracing by size features has the inherent difficulty that materials tend to be sorted with respect to settling velocity which is closely related to size. Shape of grains is not closely related to settling velocity except in extreme cases, but until recently good measurements of shape have not been available. Now the method of Fourier series grain shape description of Ehrlich & Weinberg (1970) and Ehrlich, Orzeck & Weinberg (1974) has been applied to the quartz silt sized material involved in estuarial shoaling of Charleston Harbour (Van Nieuwenhuise *et al.*, 1978). This method has allowed discrimination of the marine from fluvial components of the silts with a rough percentage estimation of the proportion of each. In the case of Charleston Harbour it is found that much of the sand involved in shoaling comes from seaward while the silt has come down-river.

Artificial tracers have mainly been used in this regard. These are mainly radioactive and are detected on the bed rather than in suspension. Important studies showing upstream movement of bed material in the lower reaches of estuaries are those of Allen & Grindley (1957) and Smith, Parsons & Cloet (1965). Where radioactive materials cannot be used, Allersma *et al.* (1973) suggest use of the natural sediment with additional trace elements, e.g. cobalt, tantalum or rare earths chemically bonded to the sediment. The material is injected, then samples are taken and the amount of the added tracer is determined by neutron activation analysis. This is tedious but may be necessary in highly populated areas. Some populated areas have peculiar tracers such as tomato seeds from sewage sludge (Shelton, 1971).

Temporal variation

There have been few studies which have made detailed analysis of the temporal variations in suspended material content, but all those that we have, show the influence of two principal variables, tidal range and water temperature. These give the tidal and seasonal effects. Some studies also find a fresh water discharge correlation. The principal workers to have examined these factors are Jackson (1964*a*, *b*) in the Humber; Halliwell & O'Connor (1967) in the Mersey; Buchan, Floodgate & Crisp (1967) in the Menai Straits; and Newton & Gray (1972) off the Yorkshire coast. The latter two regions are not strictly estuarine, but the accounts of them are the best for setting out the sources of variability in suspended sediment content.

The positive correlation of concentration with tidal range presumably reflects increased erosion of the bed under the higher shear stresses of spring tidal currents, and deposition at neaps. The seasonal variation with temperature is less clearly attributable. The effect is often a factor of four increase from summer to winter. This is more than would be expected from a change in viscosity from, say, 0.012 to 0.017 poise and it is fairly clear that temperature itself is not the controlling variable. Buchan *et al.* (1967) suggest temperature-dependent organic activity either by filter feeders trapping and pelleting the mud, or by film-forming micro-organisms trapping it in summer on tidal flats, to be responsible. The latter effect has been demonstrated by Frostick (1975), but whether it is the whole story remains to be shown. Newton & Gray (1972) on the other hand maintain that temperature is simply proxy for other physical effects with a seasonal trend, such as river discharge and wave activity. In estuaries both effects are involved. Algae trap sediment and waves resuspend it. The extent of the influence of each remains unknown.

Strategy

Finally a few notes on sampling strategy are given. Be guided by what others have found regarding spatial and temporal variability and by what is known of the physical oceanography of the estuary you are studying. At the very least the following need to be examined: neap–spring change, summer–winter change, variation along the estuary, variation across the estuary (particularly if it is well-mixed) and variation with depth. If fixed monitoring stations are to be used they should preferably be sited away from the banks of the estuary. Physical data (temperature, salinity, current speed) should be gathered at the same time, and if a coherent programme involving other workers to deal with water chemistry and organisms can be organised, so much the better. The framework provided by a numerical modelling approach is invaluable in making maximum use of the data obtained. Check and run your filtration systems in the laboratory first, and do remember to weigh the filters before you go out!

REFERENCES

Allen, F. H. & Grindley, J. (1957). Radioactive tracers in the Thames Estuary. *The Dock and Harbour Authority*, **37**, 302–6.

Allen, G. P. (1971). Déplacements saisonniers de la lentille de 'crême de vase' dans l'estuaire de la Gironde. *Comptes Rendus de l'Académie des Sciences de Paris*, **273**, 2429–31.

Allen, G. P. & Castaing, P. (1973). Suspended sediment transport from the Gironde estuary (France) onto the adjacent continental shelf. *Mar. Geol.* **14**, M47–53.

Allen, G. P., Castaing, P. & Klingbiel, A. (1974). Suspended sediment transport and deposition in the Gironde estuary and adjacent shelf. *Proceedings International Symposium on Interrelationships of Estuarine and Continental Shelf Sedimentation. Mémoires de l'Institut de Géologie du Bassin D'Aquitaine*, **7**, 27–36.

Allen, T. (1968). *Particle Size Measurement*. Chapman & Hall, London

Allersma, E., de Bruin, M., de Groot, A. J. & Houtman, J. P. W. (1973). Use of activatable tracers. In *Tracer Techniques in Sediment Transport*. International Atomic Energy Agency Technical Report **145**, Vienna.

Allersma, E., Hoekstra, A. J. & Bijker, E. W. (1967). Transport patterns in the Chao Phya estuary. *Delft Hydraulics Laboratory Publication No. 47.*

American Public Health Association (1971). *Standard methods for the examination of water and wastewater*. 13th edn, Washington, D.C.

d'Anglejan, B. F. (1970). Studies on particulate suspended material in the Gulf of St Lawrence. *McGill University, Marine Sciences Centre Manuscript Report*, **17**.

Apel, J. R. (1976). Ocean science from space. *EOS, Transactions of the American Geophysical Union*, **57**, 612–24.

Armstrong, A. C. & Brimblecombe, P. (1975). Rapid digital mapping of coastal area from LANDSAT. *Cartographic J.* **12**, 89–93.

Armstrong, A. C. & Brimblecombe, P. (1978). Experience with the use of per-point classification algorithms for the mapping of estuarine areas from LANDSAT. *J. Brit. interplanetary Soc.*, **31**, 33–6.

Baker, E. T. & Piper, D. Z. (1976). Suspended particulate matter: collection by pressure filtration and elemental analysis by thin-film X-ray fluorescence. *Deep-Sea Res.* **23**, 181–6.

Banse, K., Falls, C. P. & Hobson, L. A. (1963). A gravimetric method for determining suspended matter in sea water using Millipore filters. *Deep-Sea Res.* **10**, 639–42.

Beardsley, G. F. , Pak, H., Carder, K. & Lundgren, B. (1970). Light scattering and suspended particles in the eastern equatorial Pacific Ocean. *J. geophys. Res.* **75**, 2837–45.

Bertine, K., Carpenter, R., Cutshall, N., Robertson, D. E. & Windom, H. L. (1972). Analytical techniques for selected inorganic species. In *Marine Pollution Monitoring: Strategies for a National Program*, E. D. Goldberg

(Convenor), Workshop report, National Oceanic and Atmospheric Administration, Washington, D.C.

Bewers, J. M., Macaulay, I. D. & Sundby, B. (1974). Trace metals in the Gulf of St Lawrence. *Can. J. earth Sci.* **11**, 939–50.

Biscaye, P. E. & Eittreim, S. L. (1974). Variations in benthic boundary layer phenomena: nepheloid layer in the North American Basin. In *Suspended Solids in Water*, ed. R. J. Gibbs, pp. 227–60. Plenum Press, New York.

Brun-Cottan, J. C. (1971). Etude de la granulometrie des particules marines mesures effectuées avec un compteur Coulter. *Cahiers Oceanographiques*, **23**, 193–205.

Buchan, S., Floodgate, G. D. & Crisp, D. J. (1967). Studies on the seasonal variation of the suspended matter in the Menai Straits. I. The organic fraction. *Limnology & Oceanography*, **12**, 419–31.

Buller, A. T., Green, C. D. & McManus, J. (1975). Dynamics and sedimentation: the Tay in comparison with other estuaries. In *Nearshore Sediment Dynamics and Sedimentation*, ed. J. Hails & A. Carr, pp. 201–49. J. Wiley & Sons, London.

Cann, J. R. & Winter, C. F. (1971). X-ray fluorescence analysis of suspended sediment in sea water. *Mar. Geol.* **11**, M33–7.

Cassie, W. F., Simpson, J. R., Allen, J. H. & Hall, D. G. (1962). Final report. Hydraulic and sediment survey of the estuary of the River Tyne. *University of Durham, Kings College, Newcastle upon Tyne, Bulletin* 24.

Chester, R. & Hughes, M. J. (1967). A chemical technique for the separation of ferromanganese minerals, carbonate minerals and adsorbed trace metals from pelagic sediments. *Chem. Geol.* **2**, 249–62.

Conomos, T. J. & Gross, M. G. (1972). River–ocean suspended particulate matter relations in summer. In *The Columbia River Estuary and Adjacent Ocean waters: Bioenvironmental Studies*, ed. A. T. Pruter & D. L. Alverson, pp. 176–202. University of Washington Press, Seattle.

Conomos, T. J., Gross, M. G., Barnes, C. A. & Richards, F. A. (1972). River–ocean nutrient relations in summer. In *The Columbia River Estuary and Adjacent Ocean Waters*, ed. A. T. Pruter & D. L. Alverson, pp. 151–75. University of Washington Press, Seattle.

Conomos, T. J. & Peterson, D. H. (1974). Biological and chemical aspects of the San Francisco Bay turbidity maximum. *Proceedings International Symposium on Interrelationships of Estuarine and Continental Shelf Sedimentation. Memoires de l'Institut de Géologie du Bassin d'Aquitaine*, **7**, 45–52.

Crickmore, M. J. & Aked, R. F. (1975). Pump samplers for measuring sand transport in tidal waters. In *Instrumentation in Oceanography*, Proceedings of Institute of Electronic and Radio Engineers Conference, Bangor, Sept. 1975, 311–26.

Delft Hydraulics Laboratory (1962). *Demerara Coastal Investigation. Report on siltation of Demerara Bar Channel and Coastal erosion in British Guiana.* Delft Hydraulics Laboratory.

Di Toro, D. M., Thomann, R. V., O'Connor, D. J. & Mancini, J. L. (1977). Estuarine phytoplankton biomass models – verification analyses and preliminary applications. In *The Sea*, vol. VI, *Marine Modelling*, ed. E. D.

Goldberg, I. N. McCave, J. J. O'Brien & J. H. Steele. Wiley-Interscience, New York.

Duursma, E. K. (1967). A simple, horizontally hauled, large volume water sampling bottle of the Van Dorn type. *Deep-Sea Res.* **14**, 133–4.

Eaton, J. S., Likens, G. E. & Bormann, F. H. (1969). Use of membrane filters in gravimetric analyses of particulate matter in natural waters. *Water Resources Res.* **5**, 1151–6.

Eden, G. E. (1965). The measurement of turbidity of water, A progress report on the work of the analytical panel. *Proc. Soc. Water Treatment Examination*, **14**, 27–44.

Ehrlich, R. & Weinberg, B. (1970). An exact method for characterisation of grain shape. *J. sediment. Petrol.* **40**, 205–12.

Ehrlich, R., Orzeck, J. J. & Weinberg, B. (1974). Detrital quartz as a natural tracer–Fourier grain shape analysis. *J. sed. Petrol.* **44**, 145–50.

Flemming, G. (1970). Sediment balance of Clyde estuary. *J. Hydraulics Div. Proc. Amer. Soc. civil Eng.* **96**, 2219–30.

Folk, R. L. (1974). *Petrology of Sedimentary Rocks.* Hemphill Publishing Co., Austin, Texas.

Fowler, S. W. & Small, L. F. (1972). Sinking rates of euphausiid fecal pellets. *Limnol. Oceanogr.* **17**, 293–6.

Frostick, L. E. (1975). Sediment Studies in the Deben Estuary, Suffolk, England. Ph.D. thesis, University of East Anglia, Norwich.

Galehouse, J. S. (1971). Sedimentation analysis. In *Procedures in Sedimentary Petrology*, ed. R. E. Carver, pp. 69–94. Interscience, New York.

Gaudette, H. E. & Anderson, F. E. (1970). Possible sources of error in particulate matter analysis using silver filters. *J. sed. Petrol.* **40**, 462–4.

Gibbs, R. J. (ed.) (1974). *Suspended Solids in Water.* Plenum Press, New York.

Gibbs, R. J. (1974). Principles of studying suspended materials in water. In *Suspended Solids in Water*, ed. R. J. Gibbs, pp. 3–15. Plenum Press, New York.

Gordon, C. M. (1975). Sediment entrainment and suspension in a turbulent tidal flow. *Mar. Geol.* **18**, M57–64.

Gry, H. (1941). An apparatus for mechanical micro-analysis of sand. *Medd. fra Dansk Geol. Forening*, **10**, 17–21.

Halliwell, A. R. & O'Connor, B. A. (1967). Suspended sediment in a tidal estuary. *Proc. 10th Coastal Eng. Conf.* **1**, pp. 687–706. American Society of Civil Engineers, New York.

Harris, J. E., McKee, T. R., Wilson, R. C. & Whitehouse, U. G. (1972). Preparation of membrane filter samples for direct examination with an electron microscope. *Limnol. Oceanogr.* **17**, 784–7.

Inglis, C. C. & Allen, F. H. (1957). The regimen of the Thames estuary as affected by currents, salinities and river flow. *Proc. Inst. civil Eng.* **7**, 827–78.

Jackson, W. H. (1964*a*). Effect of tidal range, temperature and fresh water on the amount of silt in suspension in an estuary. *Nature, Lond.* **201**, 1017.

Jackson, W. H. (1964*b*). An investigation into silt in suspension in the River Humber. *The Dock and Harbour Authority*, **345**, 120–2.

Jerlov, N. G. (1976). *Marine Optics* (2nd ed. of *Optical Oceanography*). Elsevier, Amsterdam.

Jerlov, N. G. & Nielsen, E. S. (ed.) (1974). *Optical aspects of oceanography.* Academic Press, London & New York.

Jones, D. & Willis, M. S. (1956). The attenuation of light in sea and estuarine waters in relation to the concentration of suspended solid matter. *J. marine biol. Ass. UK*, **35**, 431–44.

Joseph, J. (1953). Die Trübungsverhältnisse in der Südwestlichen Nordsee Während der 'Gauss' Fahrt in Februar/März 1952. *Berichte der Deutschen Wissenschaftlichen Kommission fur Meersforschung*, **13**, 93–103.

Joseph, J. (1954). Die Sinkstoffführung von Gezeitenströmen als Austauschproblem. *Archiv für Meteorologie, Geophysik, und Bioklimatologie*, **7**, 482–501.

Joyce, J. R. (1976). Properties of fine sediments in relation to marine spoil dumping. Ph.D. thesis, University of East Anglia, Norwich.

Kajihara, M. (1971). Settling velocity and porosity of large suspended particle. *J. Oceanog. Soc. Japan*, **27**, 158–62.

Kendrick, M. P. (1972). Siltation problems in relation to the Thames barrier. *Phil. Trans. Roy. Soc.* A **272**, 223–43.

Kirby, R. & Parker, W. R. (1974). Seabed density measurements related to echo sounder records. *Dock & Harbour Authority*, **54**, 423–4.

Klemas, V., Borchardt, J. F. & Treasure, W. M. (1973). Suspended sediment observations from ERTS-12. *Remote Sensing of Environment*, **2**, 205–21.

Klemas, V., Bartlett, D., Philpot, W., Rogers, R. & Reed, L. (1974). Coastal and estuarine studies with ERTS-1 and Skylab. *Remote Sensing of Environment*, **3**, 153–74.

Kranck, K. (1973). Flocculation of suspended sediment in the sea. *Nature, Lond.*, **246**, 348–50.

Kranck, K. (1975). Sediment deposition from flocculated suspensions. *Sedimentology*, **22**, 111–23.

Krone, R. B. (1962). *Flume studies of the transport of sediment in estuarial shoaling processes.* Hydraulic Engineering Laboratory and Sanitary Engineering Research Laboratory, University of California, Berkeley.

Krone, R. B. (1963). *A Study of Rheologic Properties of Estuarine Sediments.* Sanitary Engineering Research Laboratory Report, 63–8. University of California, Berkeley.

Krone, R. B. (1972). *A Field Study of Flocculation as a Factor in Estuarial Shoaling Processes. U.S Army Corps of Engineers, Committee on Tidal Hydraulics Technical Bulletin*, **19**.

Krone, R. B. (1976). Engineering interest in the benthic boundary layer. In *The Bethic Boundary Layer*, ed. I. N. McCave, pp. 143–56. Plenum Press, New York.

Kruger, P. (1971). *Principles of Activation Analysis.* J. Wiley & Sons, New York.

Krumbein, W. C. & Pettijohn, F. J. (1938). *Manual of Sedimentary Petrography.* Appleton-Century-Crofts, New York.

Kulm, L. D. & Byrne, J. V. (1966). Sedimentary response to hydrography in an Oregon estuary. *Mar. Geol.* **4**, 85–118.

Manheim, F. T., Hathaway, J. C. & Uchupi, E. (1972). Suspended matter in surface waters of the northern Gulf of Mexico. *Limnol. Oceanog.* **17**, 17–27.

Manheim, F. T., Meade, R. H. & Bond, G. C. (1970). Suspended matter in surface waters of the Atlantic continental margin from Cape Cod to the Florida Keys. *Science*, **167**, 371–6.

Martin, J-M. & Chesselet, R. (1969). Presentation d'une nouvelle jauge nucleaire de turbidimetrie. *Comptes Rendus de la Société Géologique de la France*, **3**, 42–3.

Martin, J-M., Meybeck, M. & Heuzel, M. (1970). A study of the dynamics of suspended matter by means of natural radioactive tracers; an application to the Gironde estuary. *Sedimentology*, **14**, 27–37.

McCave, I. N. (1975). Vertical flux of particles in the ocean. *Deep-Sea Res.* **22**, 491–502.

McCave, I. N. (in press). Diagnosis of turbidites at sites 386 and 387 by particle-counter size analysis of the silt (2–40 μm) fraction. In *Initial Reports of the Deep Sea Drilling Project*, Vol. 43. U.S. Government Printing Office, Washington.

McCave, I. N. & Jarvis, J. (1973). Use of the Model T Coulter Counter in size analysis of fine to coarse sand. *Sedimentology*, **20**, 305–15.

McKenzie, K. G. (1963). The adaptation of a colorimeter for measuring silt-sized particles – a rapid photo-extinction (PE) method. *J. Sed. Petrol.* **33**, 41–8.

Meade, R. H. (1969). Landward transport of bottom sediments in estuaries of the Atlantic coastal plain. *J. Sed. Petrol.* **39**, 222–34.

Meade, R. H. (1974). Net transport of sediment through the mouths of estuaries: seaward or landward. *Proceedings International Symposium on Interrelationships of Estuarine and Continental Shelf Sedimentation. Mémoires de l'Institut de Géologie du Bassin d'Aquitaine*, **7**, 207–13.

Meade, R. H., Sachs, P. L., Manheim, F. T., Hathaway, J. C. & Spencer, D. W. (1975). Sources of suspended matter in waters of the Middle Atlantic Bight. *J. Sed. Petrol.* **45**, 171–88.

Menzel, D. W. & Vaccaro, R. F. (1964). The measurement of dissolved organic and particulate carbon in seawater. *Limnol. Oceanogr.* **9**, 138–42.

Migniot, C. (1968). Etude des propriétés physiques de différents sédiments très fins et de leur comportement sous des actions hydrodynamiques. *La Houille Blanche*, **7**, 591–620.

Newton, A. J. & Gray, J. S. (1972). Seasonal variation of the suspended solid matter, off the coast of North Yorkshire. *J. mar. biol. Ass. UK*, **52**, 33–47.

Nichols, M. M. (1974). Development of the turbidity maximum in the Rappahannock Estuary, summary, *Proceedings International Symposium on Interrelationships of Estuarine and Continental Shelf Sedimentation. Mémoires de l'Institut de Géologie du Bassin d'Aquitaine*, **7**, 19–25.

Odd, N. V. M. & Owen, M. W. (1972). A two-layer model of mud transport in the Thames Estuary. *Proc. Inst. civil Eng. Supplement*, **9**, 175–205.

Otto, L. (1966). Light attenuation in the North Sea and the Dutch Wadden Sea in relation to Secchi disc visibility and suspended matter. *Netherlands J. sea Res.* **3**, 28–51.

Owen, M. W. (1971). The effect of turbulence on the settling velocities of silt flocs. *Proc. 14th Congress int. Ass. Hydraulics Res., Paris*, Paper D-4, 27–32.

Owen, M. W. (1977). Problems in the modelling of transport, erosion, and deposition of cohesive sediments. In *The Sea VI, Marine Modelling*, ed.

E. D. Goldberg, I. N. McCave, J. J. O'Brien & J. H. Steele, pp. 515–37. Wiley-Interscience, New York.

Pak, H., Beardsley, G. F. & Park, P. K. (1970). The Columbia River as a source of marine light-scattering particles. *J. geophys. Res.* **75**, 4570–8.

Partheniades, E. (1965). Erosion and deposition of cohesive soils. *Proc. Amer. Soc. civil Eng., J. Hydraulics Division*, **91**, 105–39.

Postma, H. (1954). Hydrography of the Dutch Wadden Sea. *Archives Neerlandaises de Zoologie*, **10**, 405–511.

Postma, H. (1960). Einige Bemerkungen über den Sinkstoff transport in Ems-Dollart Gebiet. *Symposium Ems-Estuarium (Nord See). Verhandlingen van het Koninklijk Geologisch Mijnbouwkundig Genootschap*, **19**, 103–10.

Postma, H. (1961*a*). Suspended matter and Secchi disc visibility in coastal waters. *Netherlands J. Sea Res.* **1**, 359–90.

Postma, H. (1961*b*). Transport and accumulation of suspended material in the Dutch Wadden Sea. *Netherlands J. Sea Res.* **1**, 148–90.

Postma, H. (1965). Water circulation and suspended matter in Baja California lagoons. *Netherlands J. Sea Res.* **2**, 566–604.

Postma, H. (1967). Sediment transport and sedimentation in the estuarine environment. In *Estuaries*, ed. G. H. Lauff, pp. 158–79. American Association for the Advancement of Science Publication 83, Washington.

Postma, H. & Kalle, K. (1955). Die Entstehung von Trübungszonen im Unterlauf der Flüsse, speziell im Hinblick auf die Verhältnisse im der Unterelbe. *Deutschen Hydrographischen Zeitschrift*, **8**, 137–44.

Preiss, K. (1968). Non-destructive laboratory measurement of marine sediment density in a core barrel using gamma radiation. *Deep-Sea Res.* **15**, 401–7.

Price, N. B. & Calvert, S. E. (1973). A study of the geochemistry of suspended particulate matter in coastal waters. *Mar. Chem.* **1**, 169–89.

Price, N. B. & Skei, J. M. (1975). Areal and seasonal variations in the chemistry of suspended particulate matter in a deep water fjord. *Estuarine coastal mar. Sci.* **3**, 349–69.

Riley, G. A. (1970). Particulate organic matter in sea-water. *Adv. mar. Biol.* **8**, 1–118.

Schubel, J. R. (1968). Suspended sediment of the northern Chesapeake Bay. *Chesapeake Bay Institute, Johns Hopkins University, Technical Report, 35.*

Schubel, J. R. (1969). Size distributions of the suspended particles of the Chesapeake Bay turbidity maximum. *Netherlands J. Sea Res.* **4**, 283–309.

Schubel, J. R. (1971*a*). Tidal variation of the size distribution of suspended sediment at a station in the Chesapeake Bay turbidity maximum. *Netherlands J. Sea Res.* **5**, 252–66.

Schubel, J. R. (1971*b*). A few notes on the agglomeration of suspended sediment in estuaries. In *The Estuarine Environment*, ed. J. R. Schubel. American Geological Institute, Washington, D.C.

Schubel, J. R. & Schiemer, E. W. (1967). A semiautomatic microscopic particle size analyzer utilising the Vickers image splitting eyepiece. *Sedimentology*, **9**, 319–26.

Schubel, J. R. & Schiemer, E. W. (1972). A device for collecting in-situ samples of suspended sediment for microscopic analysis. *J. mar. Res.* **30**, 269–73.

Sharp, J. H. (1974). Improved analysis for 'particulate' organic carbon and nitrogen from seawater. *Limnol. Oceanogr.* **19**, 984–9.

Sheldon, R. W. (1969). A universal grade scale for particulate materials. *Proc. geol. Soc. London*, No. 1659, 293–5.

Sheldon, R. W. & Parsons, T. R. (1967). *A Practical Manual on the Use of the Coulter Counter in Marine Science*. Coulter Electronics Sales Co., Toronto.

Shelton, R. J. G. (1971). Sludge dumping in the Thames Estuary. *Mar. Pollution Bulletin*, **2**, 24–7.

Simmons, G. (1959). The photo-extinction method for the measurement of silt-sized particles. *J. Sed. Petrol.* **29**, 233–45.

Smith, D. B., Parsons, T. V. & Cloet, R. L. (1965). An investigation using radioactive tracers into the silt movement in an ebb channel, Firth of Forth, 1965. *UK Atomic Energy Research Group Report No. AERE-R5080.*

Smith, R. C., Austin, R. W. & Petzold, R. J. (1974). Volume-scattering functions in ocean waters. In *Suspended Solids in Water*, ed. R. J. Gibbs, pp. 61–72. Plenum Press, New York.

Sternberg, R. W. (1968). Friction factors in tidal channels with differing bed roughness. *Mar. Geol.* **6**, 243–60.

Strickland, J. D. H. & Parsons, T. R. (1968). *A Practical Handbook of Seawater Analysis*. Fisheries Research Board of Canada, Bulletin 167.

Subramian, V. & d'Anglejan, B. (1976). Methodology of X-ray fluorescence analysis of suspended matter in estuarine waters. *Mar. Geol.* **22**, M1–6.

Sundby, B. (1974). Distribution and transport of suspended particulate matter in the Gulf of St Lawrence. *Canad. J. earth Sci.* **11**, 1517–33.

Swift, D. J. P. & Pirie, R. G. (1970). Fine-sediment dispersal in the Gulf of San Miguel, Western Gulf of Panama: a reconnaissance. *J. mar. Res.* **28**, 69–95.

Swift, D. J. P., Schubel, J. R. & Sheldon, R. E. (1972). Size analysis of fine-grained suspended sediments: a review. *J. Sed. Petrol.* **42**, 122–34.

Terwindt, J. H. J. (1967). Mud transport in the Dutch Delta area and along the adjacent coastline. *Netherlands J. Sea Res.* **3**, 505–31.

Thiede, J., Chriss, T., Clauson, M. & Swift, S. A. (1976). Settling tubes for size analysis of fine and coarse fractions of oceanic sediments. *Oregon State University, School of Oceanography, Reference 76–8.*

Thorn, M. F. C. (1975*a*). Monitoring silt movement in suspension in a tidal estuary. *International Association for Hydraulic Research, 16th Congress, Sao Paulo*, Paper C71, pp. 596–603.

Thorn, M. F. C. (1975*b*). Deep tidal flow over a fine sand bed. *International Association for Hydraulic Research, 16th Congress, Sao Paulo*, Paper A27, pp. 217–33.

Tucholke, B. E. (1974). Determination of montmorillonite in small samples and implications for suspended matter studies. *J. Sed. Petrol.* **44**, 254–8.

Tyler, J. E. (1968). The Secchi disc. *Limnol. Oceanogr.* **13**, 1–6.

Van Nieuwenhuise, D. S., Yarus, J. M., Przygocki, R. S. & Ehrlich, R. (1978). Sources of shoaling in Charleston Harbor: Fourier grain shape analysis. *J. Sed. Petrol.* **48**, 373–83.

Vanoni, V. A. (Chairman Task Committee on preparation of sedimentation manual). (1969). Sediment measurement techniques: F. Laboratory procedures. *Proc. Amer. Soc. civil Eng., J. Hydraulics Division*, **95**, 1515–43.

Vanoni, V. A. (ed.) (1975). *Sedimentation engineering*. American Society of Civil Engineers, New York.

de Vries, M. (1970). On the accuracy of bed-material sampling. *J. hydraulic Res.* **8**, 523–33.

Whitmarsh, R. B. (1971). Precise sediment density by gamma-ray attenuation alone. *J. Sed. Petrol.* **41**, 882–3.

Williams, J. (1970). *Optical Properties in the Sea*. US Naval Institute, Annapolis.

Wollast R., Debroeu, F., Halberthal, I., Hoenig, M., Szejnberg, H., Vanderborght, J. P. & Vinikas, B. (1973). *Origine et Mècanismes de l'Envasement de l'Estuaire de l'Escaut*, Université Libre de Bruxelles, Institut de Chimie Industrielle, Rapport de Synthèse.

7

Measurement and Analysis of Water Current, Temperature, Salinity, and Density*

B. KJERFVE

Investigations into the behaviour and state of estuaries take many forms and are initiated for widely varying purposes. Field studies could include taking water samples for determination of nutrients, ATP, plankton, suspended sediments, pH, dissolved gases, and many other parameters. However, independent of the discipline, just about all field-oriented estuarine studies include some form of current velocity, water temperature, and salinity measurement.

Estuaries typically experience great spatial variations, seasonal, semi-diurnal and/or diurnal oscillations in material concentrations, water level, and flow velocity. This is primarily due to a combination of tidal influence, fresh-water discharge rate, meteorological forcing, and the constraints imposed by the configuration and geology of the estuary. Thus, the representativeness of any set of estuarine measurements is highly dependent on the sampling design (choice of sampling locations, sampling rate, and study duration) as well as a rational procedure of analysis and synthesis of the data.

It is my intention to show how estuarine velocity, salinity, and temperature data can be collected and analysed to yield both a qualitative and quantitative picture of a particular estuary. The symbols used in this discussion are listed in Table 7.1.

Sampling design

The design of an estuarine hydrographic study is critical to its meaningful completion. Variations in the measured parameters can, and usually do, occur in all three directions as well as time. However, the classical description of estuaries (summarised by Dyer, 1973) emphasises the longitudinal–vertical circulation, a two-dimensional view which typically also assumes steady state. Lately, it has become quite apparent that lateral effects may be just as important dynamically as vertical changes (Dyer, 1977). Also, it has long been

* Contribution No. 172 from the Belle W. Baruch Institute for Marine Biology and Coastal Research, University of South Carolina, Columbia, SC 29205, USA.

Table 7.1. *List of Symbols*

Symbol	Meaning
$\langle\,\rangle$	a time average
$\overline{(\,)}$	a depth average
A	area (m^2) of current cross normal to flow
B_e	equivalent bandwith of wind fluctuation (h^{-1})
c	concentration of a suspended or dissolved constituent (–)
C_D, C_1	drag coefficients (–)
$C_{T,S}, C_{15,S}, C_{15,35}\ C_{T,35}$	electrical conductivity (mmho cm^{-1})
$d(t)$	water depth (m or cm)
F	constituent flux per unit width (kg s^{-1} m^{-1})
F_T	cross-sectional constituent flux (kg s^{-1})
g	gravity (m s^{-2})
i	integer counter with respect to width
j	integer counter with respect to depth
k	integer counter with respect to time; and proportionality constant (ppt^{-1})
M	mass (kg) of current cross submerged
m	number of stations in cross section
N	number of measurements per parameter and station
n	number of sampling intervals in a tidal cycle
Q	discharge per unit with (m^2 s^{-1} or cm^2 s^{-1})
Q_T	cross-sectional discharge (m^3 s^{-1})
R_t, R_{15}	conductivity ratios (–)
S	salinity (ppt)
T	temperature (°C)
T_p	tidal period (h)
T_*	sampling duration (h)
t, t_0, t_k, t_n	time (h)
$U(z), u$	horizontal current speed (cm s^{-1})
U_*	friction velocity (cm s^{-1})
$V(\eta, t)$	horizontal velocity at η and t normal to an estuarine cross section (cm s^{-1} or m s^{-1})
W	width (m) of an estuarine sub-cross section
$X(t)$	depth-averaged triple produce (kg m^{-2} s^{-1})
$Y(\eta, t)$	the value of any parameter at η and t
z	distance (cm or m) above bottom
z_0	dynamic roughness length (cm)
α	estuarine ebb orientation (degrees)
β	coefficient used in density computations
δ	coefficient used in density computations
Δt	sampling interval (h)
$\Delta 15$	difference in conductivity ratios (–)
ε	normalised error tolerance (–)
η	non-dimensionalised measurement depth (–)
κ	Von Kàrmàn's constant
$\rho, \rho_0, \rho(t)$	water density (kg m^{-3} or g cm^{-3})
σ_T, σ_0	water density measures
τ	bed shear stress (dyne cm^{-2})
φ	angle between line and vertical (degrees)
ψ^2	mean-squared wind-driven current speed (cm^2 s^{-2})
ω	coefficient used in density computations

recognised that estuaries are seldom in steady state, which makes them difficult to treat analytically. Even the tidally averaged circulation undergoes dramatic low-frequency fluctuations in response to meteorological forcing or variations in river discharge. This was recently demonstrated by Elliott (1976) in a year-long field study in a tributary to the Chesapeake Bay.

Before making any estuarine measurements, it is wise to choose carefully a suitable sampling design. Location of stations, total number of stations, and the number of vertical measurements per station should be selected with great care and will typically vary from estuary to estuary. If discharge or fluxes of material is the goal of a study, it is usually necessary to include a minimum of three stations in a cross section. It is important to ensure that the shallow areas are adequately represented. Boon (1977) shows how the instantaneous discharge error for a cross section decreases with a greater number of stations in a section. However, if the instantaneous lateral velocity distribution is known *a priori*, a single sampling location in a cross section may suffice in determining flow discharge (Kjerfve, 1976).

In the vertical, it is usually required to make measurements at least at five locations between surface and bottom, if the water depth permits, in order that the profiles are adequately defined. Along the estuary, the station separation needs to be large enough to ensure that the differences in the mean values between the stations are larger than the sampling errors, and yet sufficiently close that the gradients between stations are reasonably linear. If there is no prior information about the dynamics of the estuary, then it will be necessary to conduct a preliminary survey to define the best station positions.

The selection of a suitable sampling duration to determine time-averaged estuarine conditions is as critical as the spatial sampling. Elliott's (1976) study points out extreme variability in time-averaged estuarine currents from one tidal cycle to another, and shows how the time-averaged current direction frequently reverses. It is likely that in most estuaries the greatest portion of the variability, on time scales from two to 20 days, occurs in response to meteorological forcing such as wind stress and atmospheric pressure fluctuations.

Weisberg (1976) argues that it is usually not sufficient to average over only a few tidal cycles to arrive at meaningful non-tidal current or flux estimates. Considering wind-induced current fluctuations, he derives an equation, which may be written

$$T_* = \frac{\Psi^2}{2B_e\varepsilon^2\langle u\rangle} \qquad (7.1)$$

and can be used to determine the necessary sampling period T_* (h). In eqn (7.1) both B_e and Ψ depend on the local wind. Ψ is the product of the wind speed (cm s^{-1}) and a non-dimensional proportionality factor equal to 0.02; B_e is an equivalent bandwidth of the local wind fluctuation with a conservative upper limit of 0.03 per hour (cph); and ε is a normalised error tolerance given as the ratio between the standard deviation and the mean of the time-averaged current, $\langle u \rangle$ (cm s^{-1}). $\langle u \rangle$ can often be estimated from previous measurements, published data, river discharge, or from conservation computations. Assume, for example, that the wind speed is 500 cm s^{-1} (such that Ψ is 10 cms^{-1}), ε is 0.2, B_e is 0.03 cph, and $\langle u \rangle$ is 10 cm s^{-1}. In this case, T_* becomes 417 h or 33 semi-diurnal tidal cycles. The main conclusion to be drawn from Weisberg's (1976) analysis is that if a too short sampling duration is selected, the resulting time-averages may not be representative of the typical conditions for a particular estuary. It should be mentioned that only a few hydrographic studies cover continuous sampling durations exceeding five tidal cycles. However in many studies the objective may be to predict the maximum concentration of a pollutant at any particular point caused by a projected discharge. In this case, the long period mean will not be of very great help. A number of measurements may be necessary and then the results will need to be treated statistically to obtain the worst likely situation.

As most surveys employ only direct reading instruments a limit of 12.5 or 25 h may be the maximum possible. In such surveys a series of stations can be occupied in sequence by a single vessel, or several stations occupied simultaneously by several vessels. As one hour is usually the longest interval between measurements that can be tolerated, only a small part of an estuary can be examined by the former method. The use of direct reading instruments also implies that measurements be taken sequentially through the water column. One must then ensure that at each level a reasonable estimate of the mean of the turbulent fluctuating value is made. This generally takes at least a minute. Even then, if the display is a visual one, some operator bias is inevitable. If measurements are made at a large number of levels the time lag between measurements needs to be taken into account in the analysis.

There are also other practical problems concerning direct measurements to be considered. The current speed in the near surface may be disturbed by the presence of the boat. Also, the current direction can be affected by the ship's magnetic field. They will be noticed and remedied by the astute observer.

Velocity measurements

Of routine estuarine measurements, current velocity is the most difficult to determine. It is a vector quantity, which usually varies greatly in both magnitude and direction relative to its resultant value over the predominant tidal period. Many current meters do not allow measurement of speed and direction simultaneously. Consequently, there are difficulties in measuring the mean speed in the mean direction. Also, any current sensor will disturb the flow somewhat and cause small deviations in the measurements from the true values.

Estuarine velocity measurements can broadly be divided into three categories. 1. Eulerian measurements, where the current velocity is determined at a point in the fluid and is recorded as a function of time. 2. Lagrangian measurements, where the path of a water parcel is followed over time. 3. Various indirect techniques, where the current velocity is averaged over a cross section.

Eulerian measurements

The Eulerian measurements have proved most successful by far in estuarine applications. This approach utilises a current meter, which can either be suspended from the surface or installed a fixed distance above the bottom. Because of the size of the sensor, the measured velocity represents a spatial average for the surface area of the sensor normal to the flow. The current speed and direction may be determined in a number of different ways.

Propeller current meter

The sensor is a propeller, often ducted, with the rate of rotation increasing with the current speed. A simple inexpensive system is described by Byrne & Boon (1973). The direction is usually sensed through a vane, which is allowed to rotate in the horizontal plane, or the meter itself swings with the current. Alternatively, the horizontal vector velocity is measured by two fixed, orthogonal speed sensors (e.g. Cannon & Pritchard, 1971). The use of propeller sensors is the most common way of measuring horizontal estuarine currents. This form of meter has excellent response characteristics. It can readily measure the fastest flow velocities which occur in estuaries, has good precision (usually $\pm 1\%$ of the full range) and threshold speeds below 10^{-2} m s^{-1}. Unfortunately, no information is gained about the vertical component velocity, which typically is two or three orders of magnitude smaller than the horizontal flow when no wind waves are present. Gordon & Dohne (1973), however, used a pivoted vane current meter which allows the current to be resolved into vertical and horizontal components. The time–mean vertical velocity can be computed numerically if several horizontal-vector meters are employed simultaneously (Pritchard, 1954).

Current cross

The sensor is a confined drag, a so-called current cross (Pritchard & Burt, 1951; Foerster, 1968), which is suspended by a line from an anchored boat. By knowing the area (A) of the cross normal to the flow, and the mass (M) of the cross while suspended in water, the current speed (V) can be determined by measuring the angle (φ) between the line and the vertical,

$$V = \left[\frac{2Mg}{A\rho C_D}\tan\varphi\right]^{\frac{1}{2}}, \qquad (7.2)$$

where g is gravity (approx. 9.81 m s^{-2}), ρ is water density (approx. 1020 kg m^{-3}), and C_D is a non-dimensional drag coefficient (approx. 1.2). By varying the mass of the current cross, the angle, φ, can be averaged by eye, usually over 10–30 s, and maintained within the desirable limits of 3° and 40°. The horizontal flow direction can be determined with a compass, and the depth of a measurement is given by the amount of line paid out multiplied by $\cos\varphi$. This technique of measuring horizontal flow vectors is inexpensive, fool-proof, and requires no power source. Depending on construction and calibration, the cross can be used to measure currents accurately and precisely with a threshold speed as low as 10^{-2} m s^{-1}. On the other hand, this technique requires people to make recordings and to take notes, and may be uncomfortable in bad weather. It is difficult to measure φ in the presence of moderate wind waves if a small boat is used, and the technique is certainly not suitable for turbulence measurements. Confined drag current meters also exist with an internal film camera, which takes photographs of an inclinometer and compass at a set sampling rate. The meter can be installed on a vertical line or be bottom-mounted. In general, confined drag current meters prove unsuitable for applications whenever wave orbital velocities are comparable in magnitude to the tidal or wind-driven current.

Savonius rotor

The sensor is a Savonius rotor, combined with a directional vane. Although this type of meter is used extensively in oceanic and fjord applications, it is not ideally suited for shallow estuaries in the presence of waves. It integrates both halves of the wave-induced water oscillation and consequently causes an over-estimate of the speed.

Acoustic doppler

The sensor consists of three acoustical (or optical) receivers. These measure the doppler-shift as a transmitted signal is scattered by particulate matter which is assumed to be moving with the ambient current (Wiseman, Crosby & Pritchard, 1972). Providing the orientation of the instrument is known, this technique has the advantage of measuring the three-dimensional velocity. Other acoustic current meters have been developed which use direct transmission between several pairs of transducers. Either the transit time of a pulse between transducers is measured or the arrival of a sound pulse is used to trigger the transmission of a further pulse. In the latter system, the frequencies of triggering are used to calculate the doppler shift and the current speed in the line of that pair of transducers. Doppler-shift instrumentation is usually too expensive and too sensitive to move from station to station to obtain vertical current profiles.

Electromagnetic flowmeters

The sensor is an electromagnetic probe which generates a magnetic field in the ambient water. Water moving through the magnetic field creates a potential difference between electrodes arranged in either one or two pairs. The voltage induced is proportional to the current velocity. This form of probe shows great promise but is still in the developmental stages with respect to estuarine current applications. Its advantages include the absence of any moving parts and the ability to measure accurately in the presence of waves. On the other hand, it is electronically complex and as a direct reading instrument it requires accurate orientation. For three-dimensional estuarine current measurements it would need to be attached to a bottom-mounted rig. These instruments are reviewed by Tucker (1972) and Kanwisher & Lawson (1975).

Other types

Other types of Eulerian flow measurements exist which are based on many different principles (Streeter, 1971, pp. 441–51), e.g. heated thermocouples, pitot tubes where the dynamic pressure may be related to velocity, and hot-wire or hot-film instrumentation with applications to turbulence. However, none of these instrument types are routinely used in hydrographic measurements.

Lagrangian measurements

Lagrangian techniques are less useful in estimating quantitative circulation features since, in narrow and shallow estuaries, a water tracer may quickly encounter boundaries. As it is impossible to recognise and follow a particular water parcel, it is necessary to tag it. This may be done through the use of dye tracers such as rhodamine (Ewart & Bendiner, 1974) and fluoroscene, or radio-isotopes. The movements of local water parcels may be determined from the spatial distribution of the tracer concentration with time. The spreading of the dye patch can be used for measurement of the turbulence characteristics of the flow. In fact, dyes are better suited for the estimation of eddy diffusivity than for the circulation features (Talbot & Talbot, 1974). Other Lagrangian current measurements which at times are applied to the study of estuaries include sea-bed drifters (Morse, Gross & Barnes, 1968), and drift bottles or other drogues (Okubo & Farlow, 1967; Okubo & Ebbesmeyer, 1976).

Indirect measurements

Geostrophic balance

At times, it may be appropriate to use indirect methods to estimate the cross-sectionally averaged flow through an estuarine section. In wide estuaries or straits, it is sometimes possible to assume that the time- and cross-sectionally averaged flow is geostrophic, i.e. there exists a balance between the cross-channel pressure gradient and the longitudinal Coriolis acceleration term (Tully, 1958). If several lateral temperature/salinity profiles exist and the surface slope due to atmospheric pressure variations and wind stress can be estimated, it is possible to compute a net longitudinal current for this section. This form of indirect current estimation may be improved by inclusion of additional acceleration terms (see Pritchard & Kent, 1956; Defant, 1961, vol. 1, pp. 513–43).

Cable voltages

A useful procedure to determine the mean instantaneous flow through a cross section involves measuring the voltage induced by a current flowing normal to a coaxial cable stretched across an estuary or sea strait. First introduced by Faraday more than a century ago, this technique has been successfully utilised by Bowden (1953) and many others, and has recently been reviewed by Sanford & Flick (1975).

Warning

As a final warning on indirect methods, it should be noted that the use of stage-discharge rating curves, so successfully applied to rivers (Morisawa, 1968), is in general not applicable to estuaries because of the oscillatory character of the predominant flow.

Determination of temperature, salinity, and density

Variations in the distribution of ocean density control the large-scale movements of water masses, are important features in the dynamics of ocean surface currents, and drive the circulation of estuaries. Density is a function of temperature, salinity, and pressure. Whereas deep-water oceanographers need to determine each of these variables, it is usually possible to neglect the pressure effects on estuarine density, since seawater for all practical purposes is incompressible in depths less than approximately 300 m.

Besides being able to express the equation of state of seawater with only two independent variables, estuarine oceanographers have another clear advantage over their blue-water colleagues. In the open ocean, temperature and salinity undergo only minor fluctuations as compared to estuaries. The coefficient of variation for temperature or salinity is typically of the order of 1% in the deep ocean, 10% in the surface waters of the open ocean, and 100% or greater in estuaries. Thus, the need for very precise temperature/salinity measurements is not as great in estuaries as in the open ocean.

Temperature

Water temperature can be measured by many means with varying accuracy and precision. Instruments commonly used include sensors based on (Besançon, 1974, p. 933). 1. Expansion of matter. 2. Changes in pressure of a gas at constant volume. 3. Thermoelectric changes due to differences in the coefficient of expansion in a bimetallic strip. 4. Changes in electrical resistance. 5. Changes in thermal radiation within a narrow wave-length band.

Mercury thermometers have long been used in oceanographic applications. They measure temperature as precisely as ±0.01 °C and are based on the thermal expansion of mercury. Specially designed reversing mercury thermometers (Duxbury, 1971, pp. 342–6) are attached to water bottles and can be used to measure/record the *in situ* temperature at any depth.

However, with the developments and improvements in electronics, thermistors have replaced mercury-filled thermometers as preferred sensors in oceanographic temperature-measuring instruments. A thermistor is a semiconductor whose electrical resistance varies inversely with temperature. A small temperature change typically causes the resistance to vary by an order of magnitude. A thermistor can be made very small, responds almost instantaneously to temperature variations, and is capable of measuring temperature with a precision of 0.001 °C. Thermistors are commonly integrated into STD (salinity–temperature–depth) sensors used for *in situ* oceanographic measurements and salinometers intended for *in situ* measurements in estuaries and coastal waters (Bowers & Coghill, 1975).

Salinity

The salinometer is now the most frequently used instrument for estuarine salinity determinations. As will be discussed in some detail below, it is an instrument with which to measure water conductivity and temperature, from which the salinity can be computed. However, the salinity can be measured by a variety of techniques. The classical procedure (Strickland & Parsons,

1968, pp. 11–19) requires a water sample which is titrated with silver nitrate. The amount of precipitable halide halogen ion from the sample is compared to the precipitate from a Standard Seawater sample. This method has a precision of approximately ± 0.01 parts per thousand (ppt) or grams solute per kilogram of seawater. For convenient, quick estimates of surface salinity to within ± 1 ppt, a refractometer is often used. It is a relatively inexpensive piece of equipment, which may be sufficient for a pre-study survey of the estuarine salinity distribution. It is based on the principle of light refracting when passing from air through a droplet of seawater whose salinity is unknown. The refractive index varies from 1.3334 for a salinity of 0 ppt to 1.3399 for 35 ppt water at 15 °C (Smith, 1974, p. 67).

However, *in situ* measurements of electrical conductivity are presently preferred in estuarine hydrographic surveys. They are quick and easy to make and yield approximately the same precision as the best titration method. As the salinity is not constant for a given conductivity but varies with the water temperature, it is essential to make temperature and conductivity measurements simultaneously.

The electrical conductivity of water is the inverse of its resistance and is expressed in units of $\text{ohm}^{-1}\ \text{cm}^{-1}$ or mho cm^{-1}. In estuarine environments the conductivity typically ranges from 10^{-3} to 50×10^{-3} mho cm^{-1} or 1–50 mmho cm^{-1}. It is either measured with an electrode system, where the conductivity cell is connected to form one arm of an a.c. Wheatstone bridge, or an inductive system, where the cell consists of two coils which are wound around an insulating core. An a.c. oscillator drives the coil, which induces a closed-loop electrical current through the core and the surrounding water. The second coil measures the induced current, which is directly proportional to the electrical conductivity of the surrounding water.

The relationship between salinity and electrical conductivity/temperature is not a simple one. Based on an extensive data set, Cox *et al.* (1967) developed an empirical computational procedure. They defined two conductivity ratios

$$R_t = C_{T,S}/C_{T,35} \quad \text{and} \quad R_{15} = C_{15,S}/C_{15,35}, \tag{7.3}$$

where $C_{T,S}$ is the measured conductivity at the *in situ* temperature, T, and salinity, S; $C_{15,S}$ is the conductivity corrected to 15 °C; and the conductivity at temperature T and salinity 35 ppt is given by

$$C_{T,35} = 29.03916(1.0 + 0.0297175T + 0.00015551T^2 - 0.000000789T^3) \tag{7.4}$$

(Perkin & Walker, 1972), which for 15 °C results in $C_{15,35} = 42.922$ mmho cm^{-1}. Cox *et al.* (1967) related R_t and R_{15} through a difference term

$$\Delta_{15} = R_{15} - R_t \tag{7.5}$$

where Δ_{15} is a function of T and R_t according to

$$\Delta_{15} = 10^{-5}R_t(R_t-1.0)(T-15)\{96.7-72.0R_t \\ +37.3R_t^2-(0.63+0.21R_t^2)(T-15)\} \quad (7.6)$$

Having computed Δ_{15}, R_{15} is found from eqn (7.5) and the salinity, S, is finally computed from

$$S = -0.08996+28.29720R_{15}+12.80832R_{15}^2-10.6789R_{15}^3 \\ +5.98624R_{15}^4-1.32311R_{15}^5 \quad (7.7)$$

This technique yields a salinity estimate with a precision of better than ± 0.01 ppt over the range 2.5–41.6 ppt. Computations were based on water samples at 1 atm pressure with temperatures from 14 °C to 29 °C.

Density

Considering time-averaged estuarine conditions, spatial salinity gradients usually correlate excellently with spatial density gradients whereas the temperature typically is almost uniform. For that reason, it is often sufficient to use salinity alone in modelling estuarine density-gradients. For example, Hansen & Rattray (1965) let

$$\rho = \rho_0(1+kS), \quad (7.8)$$

where ρ is density, ρ_0 is density of fresh water, S is salinity and k is a proportionality constant.

However, in other applications it is desirable to determine density explicitly. Although Grasshoff (1976) recounts recent studies that point to minor systematic errors in the classical formulation of the equation of state of seawater, Knudsen's (1901) method is still generally accepted for calculating density from temperature and salinity. The computational formulae are outlined below.

It is convenient to define

$$\sigma_T = (\rho-1000), \quad (7.9)$$

where ρ is density expressed in kg m^{-3}. σ_T is expressed in σ_T units, varies from approximately 1 to 30 in estuaries, and is given by

$$\sigma_T = \gamma(\sigma_0+0.1324)[1-\omega+\beta(\sigma_0-0.1324)], \quad (7.10)$$

where σ_0 is equal to σ_T at 0 °C and dependent only on the salinity, S,

$$\sigma_0 = 0.093+0.81493S-0.000482S^2+0.0000068S^3. \quad (7.11)$$

The coefficients, ω, β, and γ in eqn (7.10) are functions of temperature, T, according to

$$\omega = 10^{-3}T(4.7867 - 0.098185T + 0.0010843T^2), \tag{7.12}$$

$$\beta = 10^{-6}T(18.030 - 0.8164T + 0.01667T^2), \tag{7.13}$$

$$\gamma = -\frac{(T-3.98)^2(T+293)}{503.57(T+67.26)}. \tag{7.14}$$

Thus it is possible to compute the density or σ_T through a series of calculations, which most conveniently can be handled on a digital computer. Tables are presented in Knudsen (1901).

Shapes of measured vertical profiles

The hydrographic parameters are measured only at finite points between the surface and the bottom. These points may be more or less regularly spaced. However, to standardise the data analysis, it is desirable to interpolate the shape of each vertical profile and then read off the data values at regularly spaced depth intervals. The data points used in the analysis are therefore frequently not the measured values but rather the interpolated data.

The interpolation of the data profiles can either be accomplished graphically or numerically. However, the latter method is preferred with most data sets, as they can be analysed quickly on a computer without introducing operator errors. The most advantageous curve fit is a spline interpolation procedure. The N measured depth-values of velocity, temperature, salinity, and σ_T can be used to compute a series of cubic spline fits (Pennington, 1970, pp. 445–52; Carnahan and Wilkes, 1973, pp. 307–10). Thus, the shape of each vertical data profile is interpolated between the top-most and bottom-most data points, similarly to the manner in which a draftsman would draw a line with a flexible French curve through a sequence of points. With N depth-measurements, the spline interpolation technique requires the determination of $N-1$ piecewise continuous third order polynomials, assuming that the curvature (or second derivative) is continuous as the curve passes through each data point. A unique solution will result given suitable boundary conditions at the first and nth data points. It is usually assumed that either there is no curvature (second derivative with respect to depth equals zero) or the curvature is constant (third derivative with respect to depth equals zero) at the end points. Knowing either the slope or curvature at the end points, the profiles can be extrapolated to the surface and bottom. However, to minimise any errors introduced by the extrapolation, it is desirable that the top-most measurements are made as close to the surface as possible and that the deepest measurements preferably are made less than 1 m above the water–sediment interface. The fit of the velocity profile in the bottom boundary layer requires special considerations.

Under conditions of neutral stability and steady flow, the velocity profile close to a hydraulically rough boundary is given by the well-known logarithmic profile (Dyer, 1970; Sternberg, 1972)

$$U(z) = \frac{U_*}{\kappa} \ln (z/z_0). \tag{7.15}$$

z is the distance (cm) above the bottom. z_0 is the dynamic roughness length (cm), which represents the distance above the bottom where the velocity is zero. κ is von Kàrmàn's universal constant equal to 0.4, U_* is the friction velocity (cm s^{-1}) which is constant in the boundary layer and given by

$$U_* = (\tau/\rho)^{\frac{1}{2}}, \tag{7.16}$$

where τ is the bed shear stress (dyne cm^{-2}) and ρ is the water density, for the present purpose assumed constant and equal to 1.02 g cm^{-3}. $U(z)$ is the current speed (cm s^{-1}) a distance z above the bottom, where the speed is considered to be a short-term time-average (at least 10–30 s but often 1 min). U_* and z_0 can be determined from simultaneous current speed measurements at several heights above the boundary by regression of ln z on $U(z)$. This requires several simultaneous measurements within the bottom 1–2 m of the water column and is not available for most hydrographic studies.

Proceeding somewhat differently, it is convenient to define a non-dimensionalised drag coefficient

$$C_1 = (U_*/U_1)^2, \tag{7.17}$$

where U_1 is the current speed 1 m above the bottom. For a given boundary surface, C_1 approaches a constant value whenever the appropriate Reynolds number exceeds 10^5–1.5×10^5 which corresponds to a U_1 value in the range 10–15 cm s^{-1} (Sternberg, 1972). For most estuarine applications it seems reasonable to set $C_1 = 0.003$.

For this value of C_1 and the measured U_1, it is possible to solve for the friction velocity, U_*, from eqn (7.17). Using the computed value of U_* in eqn (7.16), the boundary shear stress may be calculated. As an example, if $U_1 = 50$ cm s^{-1}, it follows that $U_* = 2.74$ cm s^{-1} and $\tau = 7.65$ dyne cm^{-2}. A value of C_1 of 0.003 implies a fixed value of z_0 of 0.068 cm. Measurements of detailed profiles has shown that z_0 is very variable, depends greatly on the surface shape and grain size, and may range through several orders of magnitude. However, these variations do not have as large an effect on the drag coefficient.

The discussed techniques are convenient in constructing complete vertical profiles of current speed, temperature, salinity, and density from surface to bottom in estuarine environments, based on discrete vertical measurements of current speed, water temperature, and conductivity. The assumption of neutral stability in the lowest metre is usually acceptable in view of friction-induced boundary turbulence. The steady assumption necessary to assume the existence of the logarithmic current speed profile is not restrictive. The current speed varies primarily on a time-scale of the tidal period, whereas U_1 is a short-term mean on the order of 1 min.

Preliminary data manipulations

After completion of the hydrographic measurements, it is convenient to manipulate the recorded data somewhat to make possible a standardised analysis procedure. This can be accomplished using the listed computer program PROFILE (Appendix), which will be described in some detail. PROFILE uses measured values of current speed and direction, temperature and electrical conductivity at various depths as input. The program fits vertical profiles through each set of data points, computes salinity (eqns 7.3–7.7) and σ_T (eqns 7.10–7.14) profiles and gives as the final output the values of the two orthogonal horizontal velocity components, temperature, salinity, and σ_T at 11 equally spaced depths, beginning at the surface and ending at the bottom. In addition, it computes friction velocity, bottom shear stress, and dynamic roughness according to eqns (7.15–7.17) from the velocity profile. The program outputs the data in such a form that it is immediately ready to be time-averaged or otherwise manipulated and analysed.

a

HYDROGRAPHIC DATA SHEET

Name of Estuary: SOUTH SANTEE RIVER, SC
Station Number: 3
Profile Number: A69
Date: 1 FEBRUARY 1977
Time (EST): 1945
Measured Station Depth (m): 5.40

Depth (m)	Speed (cm s^{-1})	Direction	Temp. (C)	Cond. (mmho cm^{-1})
0.2	-105.	FLOOD	18.12	27.17
1.5	-121.	FLOOD	18.11	27.68
2.4	-118.	FLOOD	18.08	28.90
3.2	-98.	FLOOD	18.06	28.89
4.3	-72.	FLOOD	18.02	29.10
5.1	-61.	FLOOD	18.02	29.11

b

HYDROGRAPHIC DATA SHEET

Name of Estuary: NORTH SANTEE RIVER, SC
Station Number: 15
Profile Number: B16
Date: 1 FEBRUARY 1977
Time (EST): 0900
Measured Station Depth (m): 7.30

Depth (m)	Speed (cm s^{-1})	Direction	Temp. (C)	Cond. (mmho cm^{-1})
0.2	87.	93.	20.10	30.10
1.5	88.	89.	missing	30.23
2.5	73.	94.	20.12	30.24
3.6	68.	91.	20.05	30.71
4.4	63.	96.	20.16	31.01
5.7	50.	99.	20.19	31.04
6.8	45.	87.	20.19	31.10
7.0	36.	88.	20.19	31.10

Fig. 7.1. Two sample data sheets. It should be noted that the velocity in *a* is considered to be either ebbing (positive) or flooding (negative) such that any lateral flow is assumed negligible. In *b*, however, the velocity is expressed in terms of magnitude (always positive) and direction (degrees) in which the current is flowing.

Consider a hypothetical hydrographic study of an estuary. Velocity (speed and direction), temperature, and conductivity measurements are made at several depths between the surface and the bottom at a number of different stations. Each time a station is sampled, it is given a unique alpha-numeric identification code, time and date are noted, and the total water depth at the station is measured with a marked depth line or fathometer. As examples, consider the two sample data sheets shown in Fig. 7.1*a*, *b*.

The data at the two stations differ slightly. The velocity measurements in Fig. 7.1*a* were made assuming that the flow at all times was parallel to the estuary banks and directed either toward the ocean (ebb; positive direction) or inland (flood; negative direction). Such an assumption is usually valid in many narrow, simple estuaries. The velocity measurements in Fig. 7.1*b*, on the other hand, were made measuring both current speed and direction. The velocity output then consists of a longitudinal velocity component (positive in the main ebb-direction) and a lateral component positive 90° clockwise relative to the positive longitudinal direction. In this case it is important to select and specify the ebb-orientation ($\alpha°$) of this estuarine location relative to north. Typically, it is necessary to measure both speed and direction when the estuary is wide or has complex bathymetry and shape. Further, in Fig. 7.1*b*, it is assumed that one of the temperature measurements is missing.

The input program PROFILE consists of three types of cards. The first card, the *volume card*, specifies the number of stations which will be manipulated/analysed in a single computer run. The program has been written to allow as many as 999 stations to be treated simultaneously. Each station has an *identification card* followed by *N data cards*, where *N* is the number of depth-values of each parameter. The input cards are formated as follows.

I. *Volume Card*

Variable	*Columns*	*Format*
Number of stations, NP	3–5	I3

II. *Identification Card*

Variable	*Columns*	*Format*
Station number	1–2	I2
Profile number	6–8	A3
Estuarine orientation, $\alpha°$	10–13	F4.0
Time (military)	14–17	I4
Day	19–20	I2
Month	22–23	I2
Year (last two digits)	25–26	I2
Total station depth (m)	28–32	F5.2
Number of depth-values, *N*	34–35	I2
Check value	37	I1

Notes

(a) The estuarine orientation, α, is the angle in degrees between north and the selected ebb-direction at this station. It should only be included when the current is expressed as a speed and direction but should be omitted when the velocity is expressed as ± a speed without direction measurements in degrees.

(b) The *N*-value is the number of depth-values measured for each parameter; should one parameter have more measurements than the others, *N* is the number of depth-values for the parameter with the most measurements.

(c) The check value will vary with the type of measurements made at a particular station. If the check value is: 1, there are only velocity data; 2, there are velocity and temperature data; 3, there are velocity, temperature, and conductivity data; 4, there are temperature and conductivity data; 5, there are only temperature data. For most stations, the check value should probably be 3, i.e. velocity, temperature, and conductivity measurements were all made.

III. *N* Data Cards

Variable	*Columns*	*Format*
Station number	1–2	I2
Profile number	6–8	A3
Depth of these measurements (m)	11–20	F10.2
Velocity (cm s^{-1})	21–30	F10.0
Temperature (°C)	31–40	F10.2
Conductivity (mmho cm^{-1})	41–50	F10.2
Current direction (degrees)	51–60	F10.0

Notes

(a) If there are no current direction values, leave columns 51–60 blank; in this case, ebb-flow is denoted by positive velocities and flooding current by negative velocities. Also, be sure that columns 10–13 on the *identification card* are blank.

(b) Should some data points be missing, punch '999' in the assigned data field.

(c) If no data are available for a particular parameter, leave the assigned field blank and be sure to indicate the proper check value on the *identification card.*

(d) The *N* data cards should be arranged beginning with the measurements made closest to the surface and ending with the measurements taken closest to the bottom.

If the data from more than one station are run simultaneously (usually the case), then steps II and III are repeated up to 999 times. However, only one *volume card* is required per run.

As an example of data input to program PROFILE, Fig. 7.2 shows the correct input cards for the two stations presented in Fig. 7.1*a*, *b*. The corresponding printed output is pictured in Fig. 7.3. In addition, PROFILE is written to punch the data profiles on to computer cards (Fig. 7.4). These punched cards are formated as shown in Table 7.2.

The data values are given at equally spaced distances between surface and bottom. The non-dimensionalised depth, η, is defined

$$\eta = \frac{d(t) - z}{d(t)}, \tag{7.18}$$

where z is the distance above the bottom and d(t) is the total water depth at a given station and time (Kjerfve, 1975). For example, if the water depth is 7.30 m, the values for $\eta = 0.6$ would then represent the parameters 4.38 m below the surface. The original hydrographic measurements now exist

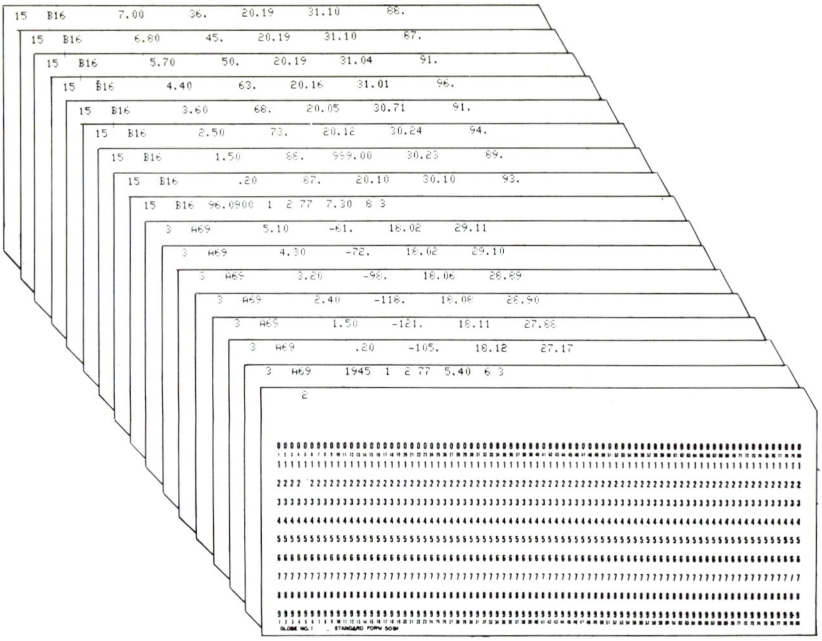

Fig. 7.2. Data input cards to PROFILE corresponding to the data in Fig. 7.1*a*, *b*.

Table 7.2. *Punched card output format from program* PROFILE

Parameter	Columns	Format					
		Velocity (cm s^{-1})	U-comp. (cm s^{-1})	V-comp. (cm s^{-1})	Temp. (°C)	Sal. (ppt)	σ_T
Station number	1–2	I2	I2	I2	I2	I2	I2
Profile number	4–6	A3	A3	A3	A3	A3	A3
Parameter id	8	V	X	Y	T	S	D
Time	10–13	I4	I4	I4	I4	I4	I4
Day	15–16	I2	I2	I2	I2	I2	I2
Month	17–18	I2	I2	I2	I2	I2	I2
Year (last two digits)	19–20	I2	I2	I2	I2	I2	I2
Depth	21–25	F5.2	F5.2	F5.2	F5.2	F5.2	F5.2
Data values at							
$\eta = 0.0$ (surface)	26–30	F5.0	F5.0	F5.0	F5.2	F5.2	F5.2
$\eta = 0.1$	31–35	F5.0	F5.0	F5.0	F5.2	F5.2	F5.2
$\eta = 0.2$	36–40	F5.0	F5.0	F5.0	F5.2	F5.2	F5.2
$\eta = 0.3$	41–45	F5.0	F5.0	F5.0	F5.2	F5.2	F5.2
$\eta = 0.4$	46–50	F5.0	F5.0	F5.0	F5.2	F5.2	F5.2
$\eta = 0.5$	51–55	F5.0	F5.0	F5.0	F5.2	F5.2	F5.2
$\eta = 0.6$	56–60	F5.0	F5.0	F5.0	F5.2	F5.2	F5.2
$\eta = 0.7$	61–65	F5.0	F5.0	F5.0	F5.2	F5.2	F5.2
$\eta = 0.8$	66–70	F5.0	F5.0	F5.0	F5.2	F5.2	F5.2
$\eta = 0.9$	71–75	F5.0	F5.0	F5.0	F5.2	F5.2	F5.2
$\eta = 1.0$ (bottom)	76–80	F5.0	F5.0	F5.0	F5.2	F5.2	F5.2

STATION NUMBER: 3 PROFILE NUMBER:A69 DATE:1945- 1- 2-1977 STATION DEPTH: 5.40 M

VELOCITY DATA (CM/SEC): POSITIVE VALUES DENOTE EBB FLOW AND NEGATIVE VALUES FLOOD FLOW

COMPUTED PARAMETERS: STRESS= 0.15E+02 DYNES/CM**2 U*= 0.38E+01 CM/SEC Z0= 0.67E-01 CM

MEASUREMENTS		COMPUTED PROFILE	
DEPTH(M)	VELOCITY	DEPTH(M)	VELOCITY
0.20	-105.00	0.0	-102.
1.50	-121.00	0.54	-110.
2.40	-118.00	1.08	-117.
3.20	-98.00	1.62	-122.
4.30	-72.00	2.16	-121.
5.10	-61.00	2.70	-112.
		3.24	-97.
		3.78	-83.
		4.32	-72.
		4.86	-64.
		5.40	0.

STATION NUMBER: 3 PROFILE NUMBER:A69 DATE:1945- 1- 2-1977 STATION DEPTH: 5.40 M

TEMPERATURE DATA (C)

MEASUREMENTS		COMPUTED PROFILE	
DEPTH(M)	TEMPERATURE	DEPTH(M)	TEMPERATURE
0.20	18.12	0.0	18.12
1.50	18.11	0.54	18.12
2.40	18.08	1.08	18.12
3.20	18.06	1.62	18.11
4.30	18.02	2.16	18.09
5.10	18.02	2.70	18.07
		3.24	18.06
		3.78	18.04
		4.32	18.02
		4.86	18.02
		5.40	18.02

STATION NUMBER: 3 PROFILE NUMBER:A69 DATE:1945- 1- 2-1977 STATION DEPTH: 5.40 M

SALINITY DATA (PPT)

MEASUREMENTS		COMPUTED PROFILE	
DEPTH(M)	SALINITY	DEPTH(M)	SALINITY
0.20	19.50	0.0	19.45
1.50	20.06	0.54	19.57
2.40	20.89	1.08	19.77
3.20	20.89	1.62	20.18
4.30	21.08	2.16	20.73
5.10	21.08	2.70	20.95
		3.24	20.89
		3.78	20.97
		4.32	21.08
		4.86	21.09
		5.40	21.08

STATION NUMBER: 3 PROFILE NUMBER:A69 DATE:1945- 1- 2-1977 STATION DEPTH: 5.40 M

DENSITY DATA (SIGMA-T UNITS)

MEASUREMENTS		COMPUTED PROFILE	
DEPTH(M)	SIGMA-T	DEPTH(M)	SIGMA-T
0.20	13.47	0.0	13.44
1.50	13.90	0.54	13.53
2.40	14.54	1.08	13.68
3.20	14.54	1.62	13.99
4.30	14.69	2.16	14.41
5.10	14.69	2.70	14.58
		3.24	14.54
		3.78	14.60
		4.32	14.70
		4.86	14.70
		5.40	14.69

STATION NUMBER:15 PROFILE NUMBER:B16 DATE: 900- 1- 2-1977 STATION DEPTH: 7.30 M

VELOCITY DATA (CM/SEC): ALPHA= 96. DEGREES: U IS POSITIVE DOWNSTREAM AND V IS POSITIVE TOWARD THE RIGHT BANK RELATIVE TO THE POSITIVE DOWNSTREAM DIRECTION

COMPUTED PARAMETERS: STRESS= 0.87E+01 DYNES/CM**2 U*= 0.29E+01 CM/SEC Z0= 0.67E-01 CM

MEASUREMENTS			COMPUTED PROFILE		
DEPTH(M)	VELOCITY	DIRECTION	DEPTH(M)	U-VALUE	V-VALUE
0.20	87.00	93.	0.0	86.	-3.
1.50	88.00	89.	0.73	89.	-10.
2.50	73.00	94.	1.46	88.	-11.
3.60	68.00	91.	2.19	77.	-4.
4.40	63.00	96.	2.92	70.	-4.
5.70	50.00	91.	3.65	68.	-6.
6.80	45.00	87.	4.38	63.	0.
7.00	36.00	88.	5.11	54.	0.
			5.84	50.	-5.
			6.57	50.	-8.
			7.30	0.	0.

```
STATION NUMBER:15      PROFILE NUMBER:B16      DATE: 900- 1- 2-1977      STATION DEPTH: 7.30 M

TEMPERATURE DATA (C)

                        MEASUREMENTS                       COMPUTED PROFILE
                        DEPTH(M)  TEMPERATURE              DEPTH(M)          TEMPERATURE
                        0.20      20.10                    0.0               20.09
                        2.50      20.12                    0.73              20.13
                        3.60      20.05                    1.46              20.15
                        4.40      20.16                    2.19              20.14
                        5.70      20.19                    2.92              20.08
                        6.80      20.19                    3.65              20.05
                        7.00      20.19                    4.38              20.16
                                                           5.11              20.20
                                                           5.84              20.19
                                                           6.57              20.19
                                                           7.30              20.19


STATION NUMBER:15      PROFILE NUMBER:B16      DATE: 900- 1- 2-1977      STATION DEPTH: 7.30 M

SALINITY DATA (PPT)

                        MEASUREMENTS                       COMPUTED PROFILE
                        DEPTH(M)  SALINITY                 DEPTH(M)          SALINITY
                        0.20      20.82                    0.0               20.82
                        1.50      20.91                    0.73              20.81
                        2.50      21.31                    1.46              20.90
                        3.60      21.48                    2.19              21.19
                        4.40      21.48                    2.92              21.42
                        5.70      21.54                    3.65              21.48
                        6.80      21.54                    4.38              21.48
                        7.00      21.54                    5.11              21.51
                                                           5.84              21.54
                                                           6.57              21.54
                                                           7.30              21.54


STATION NUMBER:15      PROFILE NUMBER:B16      DATE: 900- 1- 2-1977      STATION DEPTH: 7.30 M

DENSITY DATA (SIGMA-T UNITS)

                        MEASUREMENTS                       COMPUTED PROFILE
                        DEPTH(M)  SIGMA-T                  DEPTH(M)          SIGMA-T
                        0.20      14.01                    0.0               14.02
                        1.50      14.08                    0.73              14.00
                        2.50      14.39                    1.46              14.07
                        3.60      14.50                    2.19              14.30
                        4.40      14.50                    2.92              14.47
                        5.70      14.53                    3.65              14.50
                        6.80      14.53                    4.38              14.50
                        7.00      14.53                    5.11              14.52
                                                           5.84              14.53
                                                           6.57              14.53
                                                           7.30              14.53
```

Fig. 7.3. Sample of printed output from program PROFILE, resulting from a computer run of the input cards shown in Fig. 7.2.

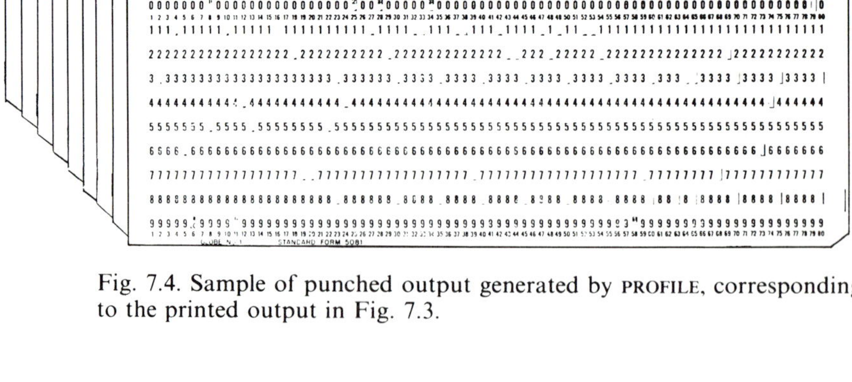

Fig. 7.4. Sample of punched output generated by PROFILE, corresponding to the printed output in Fig. 7.3.

in a form which is convenient to analyse and can be handled either by hand or by further use of a computer.

Data analysis

Time averaging

The objectives of hydrographic surveys will vary from situation to situation. The specific data analyses to be performed on the standardised instantaneous data profiles presented in the previous section will therefore vary accordingly. However, a common survey goal is usually to describe the circulation of a particular estuary and maybe to show the net movement of a dissolved or suspended constituent. In either case it is necessary to compute time-averages.

For each parameter, it is convenient to compute net or time-averaged values for the non-dimensionalised depths, $\eta = 0, 0.1, \ldots 1.0$, rather than for fixed distances below the surface or above the bottom. As discussed by Kjerfve (1975), this conserves the shape of the profiles even close to the air–sea and sea–sediment interfaces, and allows net profiles to be constructed even when the tidal range is an appreciable fraction of the mean water depth.

Preferably, measurements should be made at a constant sampling interval. Δt, over at least one complete tidal cycle. However, if at all possible, sampling should be continued for the sampling duration indicated by eqn (7.1) as net values may vary drastically from tidal cycle to tidal cycle. It is suggested that measurements begin at slack water (in most estuaries close to either low or high tide) to minimise possible averaging errors due to an erroneous choice of the duration, T_p, of a particular tidal cycle. In general

$$T_p = n\Delta t \tag{7.19}$$

where $n+1$ is the number of required profiles to compute a net profile for a given tidal cycle. Typically, in semi-diurnal or mixed tide regimes, if n is selected to be 12, the sampling rate equals 1.035 h or one *lunar hour* (Bowden, 1963), which in most cases is the longest acceptable sampling rate. Usually, it is only in retrospect that the limits of a tidal cycle can be determined, and the period is seldom an exact multiple of the sampling rate. In such cases, it is necessary to plot curves of each variable at all η-depths as functions of time and then divide each time series into n equal increments. The interpolated values at each η at the $n+1$ times would then be the data to be used in computing time averages.

The net value of one of the measured parameters, Y, at a given non-dimensionalised depth, η, is expressed by $\langle Y(\eta) \rangle$ and computed according to Kjerfve (1975)

$$\langle Y(\eta_j)\rangle = \frac{1}{n}\{\tfrac{1}{2}Y(\eta_j, t_0) + \sum_{k=1}^{n-1} Y(\eta_j, t_k) + \tfrac{1}{2}Y(\eta_j, t_n)\} \tag{7.20}$$

where $j = 0, 1, \ldots, 10$; $\eta_0, \eta_1, \ldots, \eta_{10}$ represent the non-dimensionalised depth between surface and bottom; t_0 is the starting time; t_n is the time at the end of the tidal cycle; and

$$t_k = t_0 + k\Delta t \tag{7.21}$$

for $k = 1, 2, \ldots, n-1$. The first and last profiles at t_0 and t_n, respectively are weighted by $\frac{1}{2}$, as each of these measurements is only representative of a time period $\frac{1}{2}\Delta t$. Applying eqn (7.20), net velocity, temperature, salinity, and density profiles can be computed and plotted against the non-dimensionalised depth and against the time-averaged water-depth, $\langle d\rangle$, which also is calculated using eqn (7.20).

The time-averaging of velocity must be done vectorially, which requires certain extra considerations. In many estuaries, it is sufficient to consider the flow to be parallel to the banks. The usual convention is then to consider the ebb-flow to be positive and flooding currents to be negative and time-averaging using eqn (7.20) poses no problems. However, in other estuaries, the current may show large variations in both speed and horizontal direction. It then becomes necessary to decompose the instantaneous horizontal velocity vector into two orthogonal velocity components (performed by PROFILE), and then time-average each component separately according to eqn (7.20). The resultant velocity vector is formed by combining the two net component velocities.

Net discharge computations

It is often desirable to compute the net discharge through an estuarine cross section. The velocity measurements at a station can be used to calculate the net discharge per unit cross-sectional width ($m^2\ s^{-1}$), $\langle Q\rangle$. With m stations in a section, it is then possible to integrate the m estimates of $\langle Q\rangle$ numerically and arrive at a net cross-sectional discharge, $\langle Q_T\rangle$, expressed in $m^3\ s^{-1}$.

In general, the net discharge per unit width is given by

$$\langle Q\rangle = \frac{1}{T}\int_0^T \overline{V(t)}\, \mathrm{d}(t)\, \mathrm{d}t. \tag{7.22}$$

where $\mathrm{d}(t)$ is the time-varying water depth. $\overline{V(t)}$ is the time-varying depth-averaged velocity normal to the cross section and is computed from

$$\overline{V(t)} = \tfrac{1}{10}\left\{\tfrac{1}{2}V(\eta_0, t) + \sum_{j=1}^{9} V(\eta_j, t) + \tfrac{1}{2}V(\eta_{10}, t)\right\}. \tag{7.23}$$

The surface and bottom velocities are weighted by $\frac{1}{2}$ as these boundary values are only representative over half the depth as compared to the velocity values at η_1 through η_9. Actually, it is not necessary to include $V(\eta_{10}, t)$ in eqn (7.23) as the velocity at the water–sediment interface always is zero. However, it has been included to point out that for salinity, temperature, etc., this term is non-zero. $\langle Q \rangle$ is a cross-correlation of the depth-averaged velocity and the water depth and is computed from

$$\langle Q \rangle = \frac{1}{n}\left\{\tfrac{1}{2}\overline{V(t_0)}\,\mathrm{d}(t_0) + \sum_{k=1}^{n-1} \overline{V(t_k)}\,\mathrm{d}(t_k) + \tfrac{1}{2}\overline{V(t_n)}\,\mathrm{d}(t_n)\right\}. \qquad (7.24)$$

With m stations in a cross section, the total net cross-sectional discharge $\langle Q_T \rangle$ is given by

$$\langle Q_T \rangle = \sum_{i=1}^{m} W_i \langle Q \rangle_i, \qquad (7.25)$$

where w_i is the width over which $\langle Q \rangle_i$ is assumed representative. It is usually convenient to choose the W-width in such a way that a station is located in the middle of each sub-cross section of the estuary. Further, the W's are often selected to be equally wide.

This approach makes the assumption that the estuarine cross section is rectangular. For the stations in shallow water on the sides of the estuary, the width will vary with η and with time, particularly where there are extensive inter-tidal areas. A closer approximation to the net discharge can be made assuming W to be a function only of t. In this case eqn (7.24) becomes

$$\langle Q \rangle = \frac{1}{T}\left\{\tfrac{1}{2}\overline{V(t_0)}\,\mathrm{d}(t_0)\,W(t_0) + \sum_{k=1}^{n-1} \overline{V(t_k)}\,\mathrm{d}(t_k)\,W(t_k) + \tfrac{1}{2}\overline{V(t_n)}\,\mathrm{d}(t_n)\,W(t_n)\right\} \qquad (7.26)$$

The total net cross-sectional discharge would again be given by a summation of the m stations.

The best value would be obtained by considering W also as a function of η. Then eqn (7.22) would become

$$\langle Q \rangle = \frac{1}{10}\int_0^T \int_{\eta=0}^{10} \tfrac{1}{10} V(t,\eta)\,\mathrm{d}(t)\,W(t,\eta)\,\mathrm{d}t\,\mathrm{d}\eta. \qquad (7.27)$$

One proceeds then by steps similar to eqns (7.23) and (7.24) to obtain the total net cross-sectional discharge.

Net flux computations

The goal of many biological, chemical, and geological investigations of estuaries focuses on determining the time-averaged fluxes of various dissolved or suspended constituents. As discussed by Kjerfve (1975), the net flux per unit width, $\langle F \rangle$ (kg m^{-1} s^{-1}), of an arbitrary constituent may be written

$$\langle F \rangle = \frac{1}{T}\int_0^T \overline{\rho(t)\,c(t)\,V(t)}\,\mathrm{d}(t)\,\mathrm{d}t, \qquad (7.28)$$

where $c(t)$ is the dissolved and/or suspended concentration of interest expressed as mass per mass of estuarine water. $\rho(t)$ is water density, which is related to σ_T according to eqn (7.9). In flux computations $\rho(t)$ can often be considered constant (1020 kg m^{-3}), as $\rho(t)$ varies only slightly compared to $c(t)$ and $V(t)$. The bar in eqn (7.28) represents a depth-average of the product of three instantaneous quantities and is computed from

$$\overline{\rho(t)\,c(t)\,V(t)} = \frac{1}{10}\left\{\tfrac{1}{2}\rho(\eta_0, t)\,c(\eta_0, t)\,V(\eta_0, t) + \sum_{j=1}^{9} \rho(\eta_j, t)\,c(\eta_j, t)\,V(\eta_j, t)\right\} \quad (7.29)$$

It should be noted that because $V(\eta_{10}, t) = 0$ at all times, the term representing the triple-product at the bottom vanishes and has been omitted from eqn (7.29). If $X(t) = \overline{\rho(t)\,c(t)\,V(t)}$, then the net flux per unit width, $\langle F \rangle$, is computed from

$$\langle F \rangle = \frac{1}{n}\left\{\tfrac{1}{2}X(t_0)\,\mathrm{d}(t_o) + \sum_{k=1}^{n-1} X(t_k)\,\mathrm{d}(t_k) + \tfrac{1}{2}X(t_n)\,\mathrm{d}(t_n)\right\}. \quad (7.30)$$

It probably goes without saying that if we were interested in the flux of a particular constituent, e.g. POC, nitrogen, suspended sediments, it is necessary to measure the concentration of this constituent at several depths between surface and bottom at the same time that the other measurements are made. Depth-profiles of the concentration must be constructed, and values read off at each of the η-depths for each station and sampling time. This may be rather time-consuming but necessary to compute the flux at a station. Alternatively, the curve-fitting sub-routines in the computer program PROFILE can be used.

Again, with m stations in a cross section, the total net cross-sectional flux $\langle F_T \rangle$ (kg s^{-1}) is computed from eqn (7.25) by substituting F in lieu of Q. Tidal and depth varying values of the widths assigned to each of the stations can be considered by adapting eqns (7.30) and (7.28) similarly to those introduced for eqns (7.24) and (7.22).

A numerical example

The described computations are by no means the end product of a hydrographic study. Rather, they are usually the necessary initial steps before any conclusions can be drawn about the circulation and dynamics of a particular estuary. Typically, it is consideration of the spatial distribution and variability of the various time-averages that leads to a good understanding of how an estuarine system behaves. However, that discussion is beyond the scope of this chapter.

Before closing, I will apply at least some of the averaging equations to the data set presented in Table 7.3. I will thus illustrate how to take averages

Table 7.3. *Estuarine velocity and salinity data for the 11 η-depths and 13 times, one lunar hour apart, over one tidal cycle, representative of Duplin River, GA*

		t_0	t_1	t_2	t_3	t_4	t_5	t_6	t_7	t_8	t_9	t_{10}	t_{11}	t_{12}
$d(t)$	m	6.1	6.4	6.9	7.4	7.9	8.1	8.2	8.2	7.6	7.1	6.7	6.5	6.3
$\eta = 0$	V	−2	−3	−9	−46	−60	−65	−7	73	100	89	55	24	10
	S	17.8	18.1	18.2	18.7	18.3	22.1	24.4	23.8	22.8	20.8	19.8	19.5	19.2
$\eta = 0.1$	V	−4	−6	−20	−50	−63	−65	−14	64	85	80	54	24	8
	S	17.8	18.1	18.4	18.7	19.1	22.1	24.9	27.5	23.2	20.8	19.8	19.8	19.2
$\eta = 0.2$	V	−6	−12	−33	−55	−67	−65	−29	60	83	78	63	24	2
	S	17.8	18.1	18.4	18.7	19.6	22.4	27.0	28.9	23.2	20.9	20.0	19.9	19.5
$\eta = 0.3$	V	−10	−18	−40	−55	−67	−66	−33	46	87	76	71	24	0
	S	17.8	18.2	18.5	19.0	20.6	23.2	28.0	29.5	23.2	21.0	20.1	20.0	19.7
$\eta = 0.4$	V	−12	−26	−48	−56	−66	−65	−28	40	87	73	71	22	0
	S	18.1	18.8	19.2	20.9	22.1	24.4	28.8	29.8	23.2	21.2	20.2	20.0	19.9
$\eta = 0.5$	V	−12	−29	−45	−60	−65	−64	−25	37	79	70	71	16	0
	S	18.4	20.8	21.2	21.6	23.7	26.8	29.8	30.2	24.0	21.1	20.2	20.1	20.0
$\eta = 0.6$	V	−20	−29	−40	−61	−62	−61	−22	37	73	69	71	10	0
	S	19.5	21.4	22.6	22.5	24.4	27.0	29.8	30.3	24.2	21.0	20.4	20.2	20.0
$\eta = 0.7$	V	−22	−17	−40	−60	−50	−56	−21	36	67	68	53	8	0
	S	20.2	21.9	22.7	23.4	24.8	27.1	29.8	30.3	24.2	21.0	20.4	20.2	20.0
$\eta = 0.8$	V	−20	−12	−38	−56	−50	−46	−20	36	59	68	42	5	0
	S	20.5	22.2	22.7	23.8	25.0	27.2	29.8	30.3	24.2	21.0	20.4	20.2	20.0
$\eta = 0.9$	V	−18	−8	−29	−44	−37	−34	−18	32	44	60	37	2	0
	S	20.5	22.2	22.7	23.8	25.1	27.2	29.8	30.3	24.2	21.0	20.4	20.2	20.0
$\eta = 1.0$	V	0	0	0	0	0	0	0	0	0	0	0	0	0
	S	20.5	22.2	22.7	23.8	25.1	27.2	29.8	30.3	24.2	21.0	20.4	20.2	20.0

The velocities (V) are expressed in cm s^{-1} with positive values indicating ebb flow. The salinities are given in ppt.

to compute various velocity, salinity, and depth time-averages, net discharge per unit width and net flux of salt per unit width. This data set covers one tidal cycle, with a sampling interval equal to one lunar hour, and consists of velocity (always parallel to banks) and salinity measurements. Conventional plots of these data are shown in Fig. 7.5.

Velocity averaging

Eqn (7.20) is applied at each η-depth. Referring to the velocity readings at the surface in Table 7.3, eqn (7.20) applied at η_0 becomes

$$\langle V(\eta_0)\rangle = \tfrac{1}{12}\{-\tfrac{1}{2}\times 2-3-9-46-60-65-7+73+100+89+55+24+\tfrac{1}{2}\times 10\} = 12.92 \text{ cm s}^{-1}.$$

Similarly, the net velocities at η_1 through η_{10} are : 7.58, 3.75, 1.67, −0.17, −1.75, −2.08, −1.92, −1.83, −0.33, and 0 cm s^{-1}.

The depth-averaged net velocity, $\overline{\langle V\rangle}$, at this station is computed by applying eqn (7.23) to the above time-averages:

$$\overline{\langle V\rangle} = \tfrac{1}{10}\{\tfrac{1}{2}\times 12.92+7.58+3.75+1.67-0.17-1.75-2.08-1.92-1.83-0.33\} = 1.14 \text{ cm s}^{-1} \text{ in the ebb-direction.}$$

Salinity averaging

Again, eqn (7.20) is used to compute the time-averages of salinity.

$$\langle S(\eta_0)\rangle = \tfrac{1}{12}\{\tfrac{1}{2}\times 17.8+18.1+18.2+18.7+18.3+22.1+24.4+23.8+22.8+20.8+19.8+19.5+\tfrac{1}{2}\times 19.0\} = 20.42 \text{ ppt.}$$

Similarly, the net salinities at η_1 through η_{10} are: 20.91, 21.31, 21.67, 22.30, 23.23, 23.63, 23.83, 23.92, 23.93, 23.33 ppt.

The depth-averaged net salinity, $\overline{\langle S\rangle}$, at this station is computed applying eqn (7.23) to the above time averages:

$$\overline{\langle S\rangle} = \tfrac{1}{10}\{\tfrac{1}{2}\times 20.42+20.91+21.31+21.67+22.30+23.23+23.63+23.83+23.92+23.93+\tfrac{1}{2}\times 23.93\} = 22.69 \text{ ppt.}$$

Averaging of water depth

The time-averaged water depth for this tidal cycle is computed using eqn (7.20).

$$\langle d\rangle = \tfrac{1}{12}\{\tfrac{1}{2}\times 6.1+6.4+6.9+7.4+7.9+8.1+8.2+8.2+8.2+7.6+7.1+6.7+6.5+\tfrac{1}{2}\times 6.3\} = 7.27 \text{ m.}$$

Net discharge per unit width

In computing net discharge per unit width it is first necessary to compute the depth-averaged velocity for each time. Using eqn (7.23), for $t = t_0$, this yields

$$\overline{V(t_0)} = \tfrac{1}{10}\{-\tfrac{1}{2}\times 2-4-6-10-12-12-20-22-20-18\} = -12.50 \text{ cm s}^{-1}.$$

Similarly, the depth-averaged velocity is computed for t_1 through t_n to be: -15.85, -33.75, -52.00, -55.70, -55.45, -21.35, 42.45, 71.40, 68.65, 56.05, 14.70, and 1.50 cm s^{-1}, respectively. The next step is to compute the cross correlation between the depth-averaged velocity and depth according to eqn (7.24):

$$\langle Q\rangle = \tfrac{1}{12}\{-\tfrac{1}{2}\times 12.50\times 610-15.85\times 640-33.75\times 690-\ldots.+14.70 \times 650+\tfrac{1}{2}\times 1.50\times 630\} = 270.6 \text{ cm}^2\text{ s}^{-1} \quad\text{or}\quad 0.027 \text{ m}^2\text{ s}^{-1}.$$

A note of caution is in place. It should be obvious that $\langle Q\rangle$ in general is not equal to the product of the time-averaged water depth, $\langle d\rangle$, and the net depth-averaged velocity, $\langle \overline{V}\rangle$. In this case

$$\langle d\rangle \times \langle \overline{V}\rangle = 727\times 1.14 = 828.8 \text{ cm}^2\text{ s}^{-1}.$$

The reason for this was discussed by Pritchard (1958).

Net salt flux per unit width

As an initial step in computing the net salt flux, it is necessary to calculate the density/concentration/velocity triple correlation with respect to depth according to eqn (7.29). For simplicity, I will consider density to be constant and equal to 1.02 g cm^{-3}. Salinity is expressed in ppt, whereas the concentration is non-dimensional. Thus, a salinity of 17.8 ppt corresponds to $c = 0.0178$. For t_0

$$X(t_0) = \frac{1.02}{10}\{-\tfrac{1}{2}\times 0.0178\times 2-0.0178 \times 4-0.0178\times 6-\ldots-0.0205\times 18\} = -0.13 \text{ g cm}^{-1}\text{ s}^{-1},$$

which indicates a flux of salt in the flood direction at this time. Similarly, X is evaluated at t_1 through t_n and is found to be: -0.33, -0.72, -1.13, -1.26, -1.41, -0.62, 1.24, 1.72, 1.47, 1.15, 0.30, and 0.03 g cm^{-1} s^{-1}, respectively. The net flux per unit width is calculated from eqn (7.30) accordingly:

$$\langle F\rangle = \tfrac{1}{12}\{-\tfrac{1}{2}\times 0.13\times 610-0.33\times 640-0.72\times 690-\ldots+0.30 \times 650+\tfrac{1}{2}\times 0.03\times 630\} = 6.8 \text{ g cm}^{-1}\text{ s}^{-1}$$

directed toward the ocean.

In general, the computation of net constituent flux per unit width is quite tedious. However, to save time the careful investigator does not fall into the trap of computing the flux as a product of

$$\langle d\rangle \langle \bar{c}\rangle \langle \bar{V}\rangle \rho = 727 \times 0.02269 \times 1.14 \times 1.02 = 19.8 \text{ g cm}^{-1} \text{ s}^{-1}.$$

The reasons for this difference is discussed by Pritchard (1958).

Presentation of results

Graphic presentations of the data aid greatly in the interpretation of results. The data in Table 7.3 have been displayed as time-varying velocity and salinity depth-distributions (Fig. 7.5*a*, *b*). Although generalisations about an estuary based on measurements from only one tidal cycle should be avoided, I will discuss briefly some of the data and results in this example.

The tide varies almost sinusoidally with high and low water. The symmetry of the tide curve implies a lack of pronounced river flow. High and low water coincides with slack currents, which points to a tidal wave of the standing kind (Defant, 1961, vol. 2, pp. 337–42) if friction effects are ignored. However, Hunt (1964) has shown that because of friction, water level and tidal current may be 90° out of phase, as in Fig. 7.5*a*, when the tidal wave is of the progressive kind. This matter could be resolved by analysis of several simultaneous tide and velocity records from an estuary.

The salinity, Fig. 7.5*b*, oscillates with a large amplitude over the tidal cycle. This points to discharge of a reasonable amount of fresh water to the estuary, but because of the tide curve symmetry, the estuary is not the receptacle of a major river. The short lag of the maximum salinity, 30 ppt at t_7, behind the change from flood to ebb currents is a typical phenomenon of partially mixed estuaries (Pritchard, 1952*a*).

The time-averaged profiles of velocity and salinity are shown in Fig. 7.5*c*, *d*. They indicate that the estuary indeed is partially mixed (Dyer, 1973, pp. 9–12). The net surface flow is ebb-directed whereas the bottom layer experiences a net flood flow. If the net vertical current were plotted, it would show a weak flow from the bottom to surface layer. The net salinity profile (Fig. 7.5*d*) shows a relatively weak vertical gradient without a distinctly developed halocline (pycnocline). Again, this is typical of partially mixed estuaries in the absence of major river flow.

This estuary can be classified in many different ways without knowledge of the geomorphology. Based on the shape of the net velocity and salinity profiles and the implied circulation, it is called a partially mixed or a type *B* estuary (Pritchard, 1955). Because of the net addition of fresh water, or the excess of rainfall and run-off over evaporation as indicated by the salinity plots, it has also been called a positive estuary (Pritchard, 1952*b*). However, this terminology is no longer in use as Cameron & Pritchard's (1963) commonly accepted definition of an estuary only includes positive estuaries.

a Velocity distribution (cm s^{-1}) *b* Salinity distribution ($^0/_{00}$)

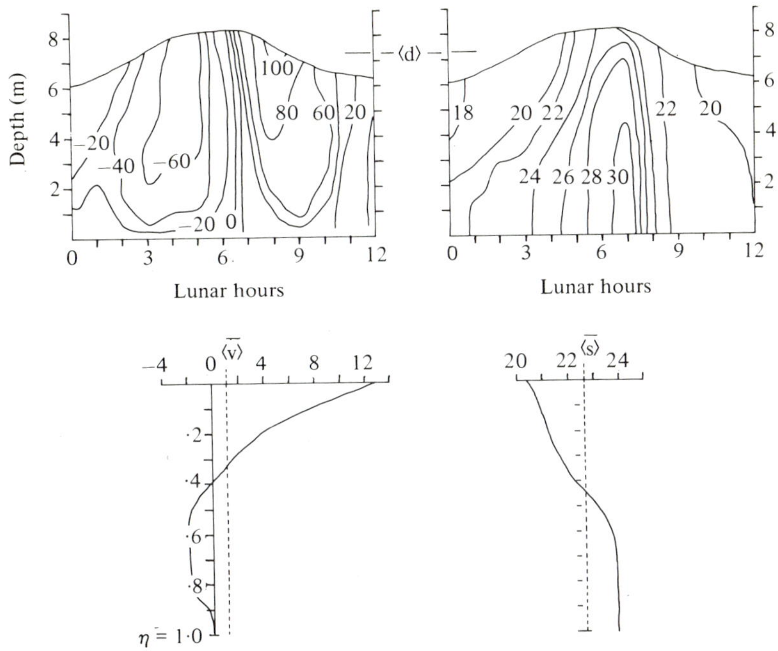

c Net velocity (cm s^{-1}) *d* Net salinity ($^0/_{00}$)

Fig. 7.5. Plots of velocity and salinity data from Table 7.3.
(*a*) Distribution of velocity at the experimental station as a function of depth and time. Positive velocity values indicate ebb flow and negative values flood flow; (*b*) Distribution of salinity against depth and time; (*c*) Net vertical velocity profile; and (*d*) Net vertical salinity profile.

By using the circulation–stratification diagram (Hansen & Rattray, 1967, Fig. 1.3), it is possible to classify the estuary in a more quantitative manner. The non-dimensional circulation parameter is computed as the net surface velocity over the depth-averaged velocity

$$\frac{\langle V(\eta_0)\rangle}{\langle V\rangle} = \frac{12.92\,(\text{cm s}^{-1})}{1.14\,(\text{cm s}^{-1})} = 11.3,$$

and implies reasonably strong gravitational circulation. The stratification parameter, also non-dimensional, is computed as the difference between net bottom and net surface salinities over the depth-averaged net salinity,

$$\frac{\langle S(\eta_{10})\rangle - \langle S(\eta_0)\rangle}{\langle S\rangle} = \frac{23.93-20.90}{22.69} = 3.0,$$

and indicates a fair degree of stratification. Because of the values of the circulation and stratification parameters, our estuary is classified as a type 2*b*. This is the type to which most estuaries belong. It indicates that our estuary dynamically is similar to e.g. the James River in the USA.

Finally, the salt flux computations per unit width indicate a small net seaward loss of salt for this tidal cycle. This is due to the slight time lag reflected in the maximum salinity occurring after the change from flooding to ebbing tide. If this trend continued, the salinity upstream of our station would have to decrease with time. However, the computed loss of 6.8 g cm^{-1} s^{-1} is very small and is likely to be reversed within the next couple of tidal cycles.

I would like to thank Ms Linda L. Vansant and Mr Sterling J. Crabtree, Jr, for their programming effort in the development of PROFILE.

REFERENCES

Besançon, R. M. (1974). Temperature and thermometry. In *The encyclopedia of physics*, 2nd edn, ed. R. M. Besançon. Van Nostrand Reinhold Co.

Boon, J. D., III. (1977). Suspended solids transport in a salt marsh creek – an analysis of errors. In *Estuarine transport processes*, ed. B. Kjerfve. University of South Carolina Press.

Bowden, K. F. (1953). Note on wind drift in a channel in the presence of tidal currents. *Proc. Roy. Soc.* A, **219**, 426–46.

Bowden, K. F. (1963). The mixing processes in a tidal estuary. *Int. J. Air and Water Pollution*, **7**, 343–56.

Bowers, R. & Coghill, G. G. (1975). A continuously profiling TSD system for use in estuarine conditions. *J. Phys. E., Scientific Instruments*, **8**, 150–2.

Byrne, R. J. & Boon, J. D. (1973). An inexpensive, fast response current speed indicator. *Chesapeake Science*, **14**, 217–19.

Cameron, W. M. & Pritchard, D. W. (1963). Estuaries. In *The Sea*, **2**, ed. M. N. Hill, pp. 306–24. Interscience Publishers, New York.

Cannon, G. A. & Pritchard, D. W. (1971). A biaxial propeller current-meter system for fixed-mount applications. *J. mar. Res.* **29**, 181–90.

Carnahan, B. & Wilkes, J. O. (1973). *Digital computing and numerical methods.* John Wiley & Sons.

Cox, R. A., Culkin, F. & Riley, J. P. (1967). The electrical conductivity/chlorinity relationship in natural sea water. *Deep-Sea Res.* **14**, 203–20.

Defant, A. (1961). *Physical oceanography.* 2 volumes. Pergamon Press, Oxford.

Duxbury, A. C. (1971). *The earth and its oceans.* Addison Wesley.

Dyer, K. R. (1970). Current velocity profiles in a tidal channel. *Geophys. J. Roy. astronom. Soc.* **22**, 153–61.

Dyer, K. R. (1973). *Estuaries: a physical introduction.* John Wiley & Sons.

Dyer, K. R. (1977). Lateral circulation effects in estuaries. In *Estuaries, geophysics and the environment.* National Academy of Sciences. Washington D.C.

Elliott, A. J. (1976). A study of the effect of meteorological forcing on the circulation in the Potomac estuary. Special Report **56**. Chesapeake Bay Institute. The Johns Hopkins University.

Ewart, T. E. & Bendiner, W. P. (1974). Techniques for estuarine and open ocean dye dispersal measurement. *Rapports et Procès-Verbaux de Réunions, Conseil International pour l'Exploration de la Mer*, **167**, 201–12.

Foerster, J. W. (1968). A portable non-electric current meter. *Chesapeake Science*, **9**, 52–5.

Gordon, C. M. & Dohne, C. F. (1973). Some observations of turbulent flow in a tidal estuary. *J. Geophys. Res.* **78**, 1971–8.

Grasshoff, K. (1976). On the possible replacement of Knudsen-Ekman's equation of state of seawater. *Deep-Sea Res.* **23**, 1079–81.

Kanwisher, J. & Lawson, K. (1975). Electromagnetic flow sensors. *Limnol. Oceanog.* **20**, 174–82.

Hansen, D. V. & Rattray, M. Jr. (1965). Gravitational circulation in straits and estuaries. *J. mar. Res.* **23**, 104–21.

Hansen, D. V. & Rattray, M. Jr. (1967). New dimensions in estuary classification. *Limnol. Oceanog.* **11**, 319–26.

Hunt, J. N. (1964). Tidal oscillations in estuaries. *Geophys. J. Roy. astronom. Soc.* **8**, 440–55.

Kjerfve, B. (1975). Velocity averaging in estuaries characterized by a large tidal range to depth ratio. *Estuarine coastal mar. Sci.* **3**, 311–23.

Kjerfve, B. (1976). Circulation and salinity distribution in coastal Louisiana bayous. *Contributions in marine Sci.* **20**, 1–10.

Knudsen, M. H. C. (1901). *Hydrographical Tables.* G. E. C. Gad, Copenhagen, 63 pp. (Reprinted in 1962. Can be obtained from G. M. Manufacturing Company 12 East St., NY).

Morisawa, M. (1968). *Streams: Their dynamics and morphology.* McGraw-Hill Book Company.

Morse, B.-A., Gross, M. G. & Barnes, C. A. (1968). Movement of seabed drifters near the Columbia River. *J. Waterways Harbors Divn, Proc. Amer. Soc. civil Eng.* **5817** (WWI), 93–103.

Okubo, A. & Ebbesmeyer, C. C. (1976). Determination of vorticity, divergence, and deformation rates from analysis of drogue observations. *Deep-Sea Res.* **23**, 349–52.

Okubo, A. & Farlow, J. S. (1967). Analysis of some Great Lakes drogue studies. *Proc. 10th Conf. Great Lakes Research. Int. Ass. Great Lakes Research, Ann Arbor*, 299–308.

Pennington, R. H. (1970). *Computer methods and numerical analysis.* 2nd edn. The Macmillan Company.

Perkin, R. G. & Walker, E. R. (1972). Salinity calculations from *in situ* measurements. *J. geophys. Res.* **77**, 6618–21.

Pritchard, D. W. (1952*a*). Salinity distribution in Chesapeake Bay. *J. mar. Res.* **6**, 106–23.

Pritchard, D. W. (1952*b*). Estuarine hydrography. In *Advances in Geophysics* **1**, 243–80.

Pritchard, D. W. (1954). A study of the salt balance in a coastal plain estuary. *J. mar. Res.* **13**, 133–44.

Pritchard, D. W. (1955). Estuarine circulation patterns. *Proc. Amer. Soc. civil Eng.* **81**, 717/1–717/11.

Pritchard, D. W. (1958). The equations of mass continuity and salt continuity in estuaries. *J. mar. Res.* **17**, 412–23.

Pritchard, D. W. & Burt, W. V. (1951). An inexpensive technique for obtaining current profiles in estuarine waters. *J. mar. Res.* **10**, 180–9.

Pritchard, D. W. & Kent, R. E. (1956). A method for determining mean longitudinal velocities in a coastal plain estuary. *J. mar. Res.* **15**, 81–91.

Sanford, T. B. & Flick, R. E. (1975). On the relationship between transport and motional electric potentials in broad, shallow currents. *J. mar. Res.* **33**, 123–9.

Smith, F. W. (ed.) (1974). *Handbook of marine science.* **1**, CRC Press.

Sternberg, R. W. (1972). Predicting initial motion and bedload transport of sediment particles in the shallow marine environment. In *Shelf sediment transport*, ed. D. J. P. Swift, D. B. Duane & O. H. Pilkey, pp. 61–82. Dowden, Hutchinson & Ross.

Streeter, V. L. (1971). *Fluid mechanics.* 5th edn. McGraw-Hill Book Company.

Strickland, J. D. H. & Parsons, T. R. (1968). A practical handbook of seawater analysis. Fisheries Research Board of Canada, Bulletin 167.

Talbot, J. W. & Talbot, G. A. (1974). Diffusion in shallow seas and in English coastal and estuarine waters. *Rapports et Procès-Verbaux de Réunions, Conseil International pour l'Exploration de la Mer*, **167**, 73–110.

Tucker, M. J. (1972). Electromagnetic current meters. *Proc. Soc. Underwater Technol.* **2**, 53–8.

Tully, J. P. (1958). On structure, entrainment and transport in estuarine enbayments. *J. mar. Res.* **17**, 523–35.

Weisberg, R. H. (1976). A note on estuarine mean flow estimation. *J. mar. Res.* **34**, 387–94.

Wiseman, W. J., Jr., Crosby, R. M. & Pritchard, D. W. (1972). A three-dimensional current meter for estuarine applications. *J. mar. Res.* **30**, 153–8.

APPENDIX

The listing of program PROFILE is presented below. It is written in FORTRAN IV, was successfully compiled using an extended H compiler, and was run on the IBM 370/168 computer system at the University of South Carolina. The test run presented in Figs. 7.2–4 required less than 52 k bytes of core storage and 6.54 s CPU time under OS/VS2.

```
C-----------------------------------------------------------------------------
C       HYDROGRAPHIC DATA COMPUTATION AND ANALYSIS
C
C           THIS PROGRAM COMPUTES ELEVEN VELOCITY,TEMPERATURE,SALINITY,AND
C       DENSITY VALUES SPACED EQUIDISTANTLY FROM SURFACE TO BOTTOM (KJERFVE,B.
C       1975.'VELOCITY AVERAGING IN ESTUARIES CHARACTERIZED BY A LARGE TIDAL
C       RANGE TO DEPTH RATIO'.ESTUARINE AND COASTAL MARINE SCIENCE 3:311-323).
C       SALINITY AND DENSITY (SIGMA-T) ARE COMPUTED FROM TEMPERATURE AND
C       CONDUCTIVITY MEASUREMENTS (COX,R.A.,F.CULKIN, AND J.P.RILEY. 1967.
C       'THE ELECTRICAL CONDUCTIVITY/CHLORINITY RELATIONSHIP IN NATURAL SEA WATER'
C       DEEP-SEA RESEARCH 14:203-220).
C       A CUBIC SPLINE INTERPOLATION IS USED TO FIT A CURVE THROUGH THE
C       PLOTTED VALUES VERSUS DEPTH (PENNINGTON, R.H. 1970. 'INTRODUCTORY
C       COMPUTER METHODS AND NUMERICAL ANALYSIS.' SECOND EDITION. THE MCMILLAN
C       COMPANY: 445-452).
C       A LOGARITHMIC CURVE IS FITTED FOR THE VELOCITY FROM 1 M ABOVE THE
C       BOTTOM TO THE BOTTOM.
C-----------------------------------------------------------------------------
C       DEFINITION OF VARIABLES
C           A. INPUT VARIABLES, INTEGERS
C                 CHECK: IF EQUAL TO:      READ IN:
C                                 1        VELOCITY DATA
C                                 2        VELOCITY AND TEMPERATURE DATA
C                                 3        VELOCITY,TEMPERATURE AND CONDUCTIVITY DATA
C                                 4        TEMPERATURE AND CONDUCTIVITY DATA
C                                 5        TEMPERATURE DATA
C                 DAY: DAY OF MONTH, FORMAT I2
C                 MONTH: MONTH OF YEAR, FORMAT I2
C                 N: NUMBER OF DATA POINTS PER PROFILE (NO MORE THAN 20), FORMAT I2
C                 NP: NUMBER OF PROFILES (LESS THAN 1000), FORMAT I3
C                 PROF: PROFILE DISIGNATION, FORMAT A3
C                 SNUM: STATION NUMBER (LESS THAN 100), FORMAT I2
C                 TIME: 24-HOUR CLOCK, FORMAT I4
C                 YEAR: FORMAT I2
C           B. INPUT VARIABLES, REAL
C                 ALPHA: THE DIRECTION ANGLE IN DEGREES BETWEEN TRUE NORTH
C                        AND THE LOCAL DOWNSTREAM DIRECTION (X-AXIS), FORMAT F5.0 *
C                        NOTE: IF DIRECTION IS NOT MEASURED AND THE CURRENT IS
C                        CONSIDERED TO BE EITHER EBBING OR FLOODING, ALPHA SHOULD
C                        BE OMITTED.
C                 COND: ELECTRICAL CONDUCTIVITY (MMHO/CM), FORMAT F5.2  *
C                 DEP: DEPTH (M) AT WHICH MEASUREMENT IS MADE, FORMAT F5.2
C                 DEPTH: STATION DEPTH (M),   FORMAT F5.2
C                 OMEGA: DOWNSTREAM CURRENT DIRECTION IN DEGREES, FORMAT F5.0   *
C                 TEMP: TEMPERATURE (C), FORMAT F5.2  *
C                 VEL: CURRENT SPEED (CM/SEC): +:EBB,-:FLOOD, FORMAT F5.0 *
C
C                 *  VALUES MUST BE LESS THAN 999.00
C           C. SELECTED INTERNAL VARIABLES
C                 BETA: DIFFERENCE IN DEGREES BETWEEN OMEGA AND ALPHA
C                 DATA: VALUES OF THE MEASURED OR COMPUTED PARAMETERS
C                 ID: 1=OMEGA VALUES
C                     2=VELOCITY VALUES
C                     3=TEMPERATURE VALUES
C                     4=SALINITY VALUES
C                     5=DENSITY VALUES
C                 ID2: DA=DENSITY VALUES
C                      S=SALINITY VALUES
C                      T=TEMPERATURE VALUES
C                      X=LONGITUDINAL VELOCITY COMPONENT
C                      Y=LATERAL VELOCITY COMPONENT
C                      V=VELOCITY VALUES WHENEVER ALPHA EQUALS ZERO
```

```
C               PV: THE CALCULATED VALUES SPACED EQUIDISTANTLY FROM SURFACE TO
C                   BOTTOM
C               RHO: WATER DENSITY, ASSUMED TO EQUAL 1.02 G/CC
C               SAL: SALINITY (PPT)
C               SIGT: DENSITY (SIGMA-T UNITS)
C               STRESS: BOUNDARY SHEAR STRESS (DYNES/CM**2)
C               USTAR: FRICTION VELOCITY (CM/SEC)
C               UVEC: U COMPONENT OF DECOMPOSED CURRENT VELOCITY VECTOR
C               VEL1: CALCULATED VALUES OF OMEGA SPACED EQUIDISTANTLY FROM
C                     SURFACE TO BOTTOM
C               VVEC: V COMPONENT OF DECOMPOSED CURRENT VELOCITY VECTOR
C               ZNOT: ROUGHNESS LENGTH (CM)
C-----------------------------------------------------------------------------
      DIMENSION DEP(21),DATA(21),SIGT(21),SAL(21),COND(21),VEL(21),D(12)
     *,PV(12),TEMP(21),OMEGA(21),VEL1(12),UVEC(12),VVEC(12),DEP1(21)
      DIMENSION C(4,21)
      INTEGER SNUM,PROF,CHECK,TIME,DAY,YEAR,COUNT,A,B
      INTEGER DA,S,T,X,Y,V
      DATA DA/'D'/,S/'S'/,T/'T'/,X/'X'/,Y/'Y'/,V/'V'/
      INREAD=5
      IPRINT=6
      IPUNCH=7
      READ(INREAD,900) NP
  900 FORMAT(I5,75X)
      DO 1010 I10=1,NP
      READ(INREAD,902) SNUM,PROF,ALPHA,TIME,DAY,MONTH,YEAR,DEPTH,N,
     *CHECK
  902 FORMAT(I2,3X,A3,F5.0,I4,1X,I2,1X,I2,1X,I2,1X,F5.2,1X,I2,1X,I1)
      GO TO (100,110,120,130,140),CHECK
  100 IF(ALPHA.EQ.0.) GO TO 105
      READ(INREAD,904)(DEP(I),VEL(I),OMEGA(I),I=1,N)
  904 FORMAT(10X,F10.2,F10.0,20X,F10.0,20X)
      ID=1
      GO TO 150
  105 READ(INREAD,906)(DEP(I),VEL(I),I=1,N)
  906 FORMAT(10X,F10.2,F10.0,50X)
      ID=2
      GO TO 150
  110 IF(ALPHA.EQ.0.) GO TO 115
      READ(INREAD,908)(DEP(I),VEL(I),TEMP(I),OMEGA(I),I=1,N)
  908 FORMAT(10X,F10.2,F10.0,F10.2,10X,F10.0,20X)
      ID=1
      GO TO 150
  115 READ(INREAD,910)(DEP(I),VEL(I),TEMP(I),I=1,N)
  910 FORMAT(10X,F10.2,F10.0,F10.2,40X)
      ID=2
      GO TO 150
  120 IF(ALPHA.EQ.0.) GO TO 125
      READ(INREAD,912)(DEP(I),VEL(I),TEMP(I),COND(I),OMEGA(I),I=1,N)
  912 FORMAT(10X,F10.2,F10.0,2F10.2,F10.0,20X)
      ID=1
      GO TO 150
  125 READ(INREAD,914)(DEP(I),VEL(I),TEMP(I),COND(I),I=1,N)
  914 FORMAT(10X,F10.2,F10.0,2F10.2,30X)
      ID=2
      GO TO 150
  130 READ(INREAD,916)(DEP(I),TEMP(I),COND(I),I=1,N)
  916 FORMAT(10X,F10.2,10X,2F10.2,30X)
      ID=3
      GO TO 150
  140 READ(INREAD,918)(DEP(I),TEMP(I),I=1,N)
  918 FORMAT(10X,F10.2,10X,F10.2,40X)
      ID=3
  150 IF(DEP(N).GT.DEPTH) GO TO 550
      IF(ID.NE.2) GO TO 155
      IF(DEP(N).EQ.DEPTH) GO TO 560
  155 GO TO (160,160,170,170,160),CHECK
  170 CALL SALT(COND,TEMP,SAL,SIGT,N)
  160 DINC=DEPTH/10.
      D1=DEPTH-1.0
      D(1)=0.0
      I21=0
```

```
      DO 1020 I20=2,11
      IF(I20.EQ.I21) GO TO 1020
      I19=I20-1
      D(I20)=D(I19)+DINC
      IF(D1.LT.D(I19).OR.D1.GT.D(I20)) GO TO 1020
      I21=I20+1
      D(I21)=D(I20)
      D(I20)=D1
      L1=I20
 1020 CONTINUE
      D(12)=DEPTH
      COUNT=0
      DO 1025 I25=1,N
      DEP1(I25)=DEP(I25)
 1025 CONTINUE
      I120 = 0
      IF (D (1) .EQ. DEP (1)) GO TO 2020
      DO 2000 LC11 = 1,11
      IF (D (LC11) .GE. DEP (1)) GO TO 2010
 2000 CONTINUE
 2010 I120 = LC11 - 1
 2020 I122 = 12
      I123 = 12
      I121 = 12
      IF (D (12) .EQ. DEP (N)) GO TO 2090
      DO 2080 LC12 = 1,11
      IF (D (I121) .LE. DEP (N)) GO TO 2090
      I121 = I121 - 1
 2080 CONTINUE
 2090 CONTINUE
      IH121 = I121
      IH123 = I123
  180 DO 1030 I30=1,12
      PV(I30)=0.0
 1030 CONTINUE
      I121 = IH121
      I123 = IH123
      IF(ID.EQ.2) GO TO 185
      DO 1035 I35=1,N
      DEP(I35)=DEP1(I35)
 1035 CONTINUE
  185 GO TO (190,200,210,220,186),ID
  186 DO 1040 I40=1,N
      DATA(I40)=SIGT(I40)
 1040 CONTINUE
      ID2=DA
      GO TO 250
  190 DO 1050 I50=1,N
      DATA(I50)=OMEGA(I50)
 1050 CONTINUE
      GO TO 250
  200 DO 1060 I60=1,N
      DATA(I60)=VEL(I60)
 1060 CONTINUE
      ID2=X
      GO TO 250
  210 DO 1070 I70=1,N
      DATA(I70)=TEMP(I70)
 1070 CONTINUE
      ID2=T
      GO TO 250
  220 DO 1080 I80=1,N
      DATA(I80)=SAL(I80)
 1080 CONTINUE
      ID2=S
  250 NK=0
      N1=N
```

```
      NC=N
      I100=1
  252 IF(DATA(I100).NE.999.) GO TO 254
      NJ=NC-1
      DO 1110 I110=I100,NJ
      DATA(I110)=DATA(I110+1)
      DEP(I110)=DEP(I110+1)
      GO TO (253,255,258,1110,1110),ID
      GO TO 1110
  253 VEL(I110)=VEL(I110+1)
  255 OMEGA(I110)=OMEGA(I110+1)
      GO TO 1110
  258 SAL(I110)=SAL(I110+1)
      SIGT(I110)=SIGT(I110+1)
 1110 CONTINUE
      NC=NC-1
      GO TO 252
  254 IF(I100.GE.NC)GO TO 256
      I100=I100+1
      GO TO 252
  256 NK=N-NC
      IF(ID.EQ.2.AND.NK1.NE.0) NC=NC-NK1
      IF(NC.EQ.1) GO TO 570
      GO TO (230,257,257,230,230),ID
  230 DO 1090 I90=2,NC
      IX=0
      EQ1=DATA(I90)-DATA(I90-1)
      IF(ID.EQ.1) GO TO 240
      EQ1=ABS(EQ1)
      IF(EQ1.LE.0.01) DATA(I90)=DATA(I90-1)
      GO TO 1090
  240 IF(EQ1.GE.90.) GO TO 245
      IF(EQ1.LE.-90.) DATA(I90)=DATA(I90)+360.
      GO TO 1090
  245 DATA(I90-1)=DATA(I90-1)+360.
      IX=1
 1090 CONTINUE
      IF(IX.EQ.1) GO TO 230
  257 CONTINUE
      LZ = NC
      IF (DEP (NC) .EQ. D (I123)) GO TO 2340
      IF (ID .NE. 2) GO TO 2300
      IF (I121 .GE. L1) GO TO 2340
 2300 CONTINUE
      CALL OUTSD (DEP (NC), DATA (NC), DEP (NC-1), DATA (NC-1), SM)
      LZ = NC + 1
      DEP (NC+1) = DEPTH
      DATA (LZ) = SM * (DEP (NC+1) - DEP (NC)) + DATA (NC)
 2340 CONTINUE
      CALL SPLIC (DEP, DATA, LZ, C)
      I119 = I120 + 1
      IF (ID .NE. 2) GO TO 2120
      I123 = L1
 2120 DO 2140 LC14 = I119,I123
      CALL SPLIN (DEP, DATA, LZ, C, D (LC14), PV (LC14))
 2140 CONTINUE
      IF (I120.EQ.0) GO TO 2105
      IF (DEP (1) .EQ. D (I119)) I119=I119+1
      CALL OUTSD (DEP(1),DATA(1),D(I119),PV(I119),SM)
      DO 2100 LC13 = 1,I120
      PV (LC13) = SM * (D (LC13) - DEP (1)) + DATA (1)
 2100 CONTINUE
 2105 CONTINUE
      IF (ID .EQ. 2) GO TO 2230
      GO TO 330
 2230 CONTINUE
  320 CALL LOGFIT (DEPTH,D,DEP,DATA,LZ,PV,STRESS,USTAR,ZNOT,L1)
```

```
330 COUNT=COUNT+1
    IF(ID.NE.1) GO TO 335
    IF(ALPHA.EQ.0.) GO TO 335
    GO TO 500
335 IF(ID.EQ.2) GO TO 336
    WRITE(IPRINT,920) SNUM,PROF,TIME,DAY,MONTH,YEAR,DEPTH
920 FORMAT('0',5X,'STATION NUMBER:',I2,5X,'PROFILE NUMBER:',A3,5X,'DAT
   *E:',I4,'-',I2,'-',I2,'-19',I2,5X,'STATION DEPTH:',F5.2,' M')
    GO TO 337
336 WRITE(IPRINT,921) SNUM,PROF,TIME,DAY,MONTH,YEAR,DEPTH
921 FORMAT('1',5X,'STATION NUMBER:',I2,5X,'PROFILE NUMBER:',A3,5X,'DAT
   *E:',I4,'-',I2,'-',I2,'-19',I2,5X,'STATION DEPTH:',F5.2,' M')
337 GO TO (500,350,370,380),ID
    WRITE(IPRINT,922)
922 FORMAT('0',5X,'DENSITY DATA (SIGMA-T UNITS)')
    GO TO 400
350 IF(ALPHA.EQ.0.) GO TO 360
    CALL VECTOR(PV,VEL1,L1,UVEC,VVEC,ALPHA)
    WRITE(IPRINT,924)  ALPHA
924 FORMAT('0',5X,'VELOCITY DATA (CM/SEC):  ALPHA=',F4.0,' DEGREES:  U
   * IS POSITIVE DOWNSTREAM AND V IS POSITIVE TOWARD THE RIGHT BANK RE
   *LATIVE',/,31X,'TO THE POSITIVE DOWNSTREAM DIRECTION')
    GO TO 400
360 WRITE(IPRINT,926)
926 FORMAT('0',5X,'VELOCITY DATA (CM/SEC):  POSITIVE VALUES DENOTE EBB
   * FLOW AND NEGATIVE VALUES FLOOD FLOW')
    GO TO 400
370 WRITE(IPRINT,928)
928 FORMAT('0',5X,'TEMPERATURE DATA (C)')
    GO TO 400
380 WRITE(IPRINT,930)
930 FORMAT('0',5X,'SALINITY DATA (PPT)')
400 IF(ID.NE.2) GO TO 420
    IF(STRESS.EQ.0.) GO TO 410
    WRITE(IPRINT,932) STRESS,USTAR,ZNOT
932 FORMAT('0',10X,'COMPUTED PARAMETERS:  STRESS=',E9.2,' DYNES/CM**2'
   *,5X,'U*=',E9.2,' CM/SEC',5X,'Z0=',E9.2,' CM')
    GO TO 420
410 WRITE(IPRINT,934) STRESS,USTAR
934 FORMAT('0',10X,'COMPUTED PARAMETERS:  STRESS=',E9.2,' DYNES/CM**2'
   *,5X,'U*=',E9.2,' CM/SEC',5X,'Z0=----')
420 WRITE(IPRINT,936)
936 FORMAT('0',36X,'MEASUREMENTS',20X,'COMPUTED PROFILE')
    IF(ID.NE.2) GO TO 422
    IF(ALPHA.NE.0.) GO TO 425
422 GO TO (500,432,431,433,423),ID
423 WRITE(IPRINT,937)
937 FORMAT(' ',36X,'DEPTH(M)',1X,'SIGMA-T',16X,'DEPTH(M)',8X,'SIGMA-T'
   *)
    GO TO 426
432 WRITE(IPRINT,943)
943 FORMAT(' ',36X,'DEPTH(M)',1X,'VELOCITY',15X,'DEPTH(M)',7X,'VELOCIT
   *Y')
    GO TO 426
431 WRITE(IPRINT,938)
938 FORMAT(' ',36X,'DEPTH(M)',1X,'TEMPERATURE',12X,'DEPTH(M)',8X,'TEMP
   *ERATURE')
    GO TO 426
433 WRITE(IPRINT,947)
947 FORMAT(' ',36X,'DEPTH(M)',1X,'SALINITY',15X,'DEPTH(M)',8X,'SALINIT
   *Y')
    GO TO 426
425 WRITE(IPRINT,939)
939 FORMAT(' ',34X,'DEPTH(M)',3X,'VELOCITY',1X,'DIRECTION',6X,'DEPTH(M
   *)',8X,'U-VALUE',5X,'V-VALUE')
426 DO 1160 I160=1,10
    IF(PV(I160).LT.0.0.AND.PV(I160).GT.-1.0) PV(I160)=0.0
```

```
1160 CONTINUE
     IF(ID.EQ.2) GO TO 430
     DO 1140 I140=1,NC
     IF(I140.EQ.L1) GO TO 1140
     WRITE(IPRINT,940) DEP(I140),DATA(I140),D(I140),PV(I140)
 940 FORMAT(' ',34X,F6.2,3X,F7.2,17X,F5.2,12X,F5.2)
1140 CONTINUE
     GO TO 436
 430 IF(ALPHA.EQ.0.) GO TO 435
     IF(NCC.LT.NC) NC=NCC
     DO 1141 I141=1,NC
     WRITE(IPRINT,941)DEP(I141),DATA(I141),OMEGA(I141),D(I141),
    *UVEC(I141),VVEC(I141)
 941 FORMAT(' ',34X,F6.2,3X,F7.2,4X,F5.0,8X,F5.2,11X,F5.0,8X,F5.0)
1141 CONTINUE
     GO TO 436
 435 DO 1142 I142=1,NC
     WRITE(IPRINT,942) DEP(I142),DATA(I142),D(I142),PV(I142)
 942 FORMAT(' ',34X,F6.2,5X,F7.2,15X,F5.2,11X,F5.0)
1142 CONTINUE
 436 M=NC+1
     DO 1150 I150=M,11
     IF(I150.EQ.L1) GO TO 1150
     IF(ID.EQ.2) GO TO 440
     WRITE(IPRINT,944) D(I150),PV(I150)
 944 FORMAT(' ',67X,F5.2,12X,F5.2)
     GO TO 1150
 440 IF(ALPHA.EQ.0.) GO TO 445
     WRITE(IPRINT,945) D(I150),UVEC(I150),VVEC(I150)
 945 FORMAT(' ',67X,F5.2,11X,F5.0,8X,F5.0)
     GO TO 1150
 445 WRITE(IPRINT,946) D(I150),PV(I150)
 946 FORMAT(' ',67X,F5.2,11X,F5.0)
1150 CONTINUE
     IF(ID.EQ.2) GO TO 450
     WRITE(IPRINT,948) DEPTH,PV(12)
 948 FORMAT(' ',67X,F5.2,12X,F5.2,//)
     GO TO 460
 450 IF(ALPHA.EQ.0.) GO TO 455
     WRITE(IPRINT,949) DEPTH,UVEC(12),VVEC(12)
 949 FORMAT(' ',67X,F5.2,11X,F5.0,8X,F5.0,//)
     GO TO 460
 455 WRITE(IPRINT,950) DEPTH,PV(12)
 950 FORMAT(' ',67X,F5.2,11X,F5.0,//)
 460 XDEP = DEPTH
     NC1=L1+1
     NQ=L1-1
     IF(ID.EQ.2) GO TO 465
     WRITE(IPUNCH,962) SNUM,PROF,ID2,TIME,DAY,MONTH,YEAR,XDEP,(PV(J),
    *J=1,NQ),(PV(J),J=NC1,12)
 962 FORMAT(I2,1X,A3,1X,A1,1X,I4,1X,3I2,F5.2,11F5.2)
     GO TO 468
 465 IF(ALPHA.EQ.0.) GO TO 466
     ID2=X
     WRITE(IPUNCH,964) SNUM,PROF,ID2,TIME,DAY,MONTH,YEAR,DEPTH,(UVEC(J)
    *,J=1,NQ),(UVEC(J),J=NC1,12)
     ID2=Y
     WRITE(IPUNCH,964) SNUM,PROF,ID2,TIME,DAY,MONTH,YEAR,DEPTH,
    *(VVEC(J),J=1,NQ),(VVEC(J),J=NC1,12)
     GO TO 468
 964 FORMAT(I2,1X,A3,1X,A1,1X,I4,1X,3I2,F5.2,11F5.0)
 466 ID2=V
     WRITE(IPUNCH,964) SNUM,PROF,ID2,TIME,DAY,MONTH,YEAR,DEPTH,(PV(J),J
    * =1,NQ),(PV(J),J=NC1,12)
     GO TO 468
 468 GO TO (1010,470,480,490,1010),CHECK
 470 GO TO (505,1010,480,480,480),COUNT
```

```
  480 GO TO (505,510,520,1010,490),COUNT
  490 GO TO (510,520,1010,500,500),COUNT
  500 ID=2
      NCC=NC
      NK1=NK
      COUNT=COUNT-1
      DO 1170 I170=1,12
      VEL1(I170)=PV(I170)
 1170 CONTINUE
      GO TO 180
  505 ID=3
      GO TO 180
  510 ID=4
      GO TO 180
  520 ID=5
      GO TO 180
  550 WRITE(IPRINT,952) PROF
  952 FORMAT('0',5X,'ERROR IN PROFILE NUMBER:',A4,' THE LAST PROFILE DEP
     *TH IS GREATER THAN THE TOTAL DEPTH !!!',////)
      GO TO 1010
  560 WRITE(IPRINT,954) PROF
  954 FORMAT('0',5X,'ERROR IN PROFILE NUMBER:',A4,' THE LAST PROFILE DEP
     *TH FOR VELOCITY DATA MUST BE LESS THAN THE TOTAL DEPTH !!!',////)
      GO TO 1010
  570 WRITE(IPRINT,956) PROF
  956 FORMAT('0',5X,'ERROR IN PROFILE NUMBER:',A4,'  ONE OF THE PROFILE
     *PARAMETERS CONTAINS ONLY ONE VALID DATA POINT!!',/,5X,'A PROFILE C
     *ANNOT BE COMPUTED FOR THAT PARAMETER!!!!')
 1010 CONTINUE
      STOP
      END
```

```
C-----------------------------------------------------------------------
C     SUBROUTINE: LOGARITHMIC FIT
C     THIS SUBROUTINE FITS A LOGARITHMIC CURVE FROM 1 M ABOVE THE BOTTOM
C     TO THE BOTTOM.  STRESS COMPUTATIONS ARE MADE ACCORDING TO
C     STERNBERG,R.W. 1972. 'PREDICTING INITIAL MOTION AND BEDLOAD TRANSPORT OF
C     SEDIMENT PARTICLES IN THE SHALLOW MARINE ENVIRONMENT'. SHELF SEDIMENT
C     TRANSPORT. D.J.P.SWIFT,D.B.DUANE, AND O.H.PILKEY,EDS. DOWDEN, HUTCHINSON
C     AND ROSS:61-82.
C-----------------------------------------------------------------------
      SUBROUTINE LOGFIT(DEPTH,D,DEP,DATA,N,PV,STRESS,USTAR,ZNOT,L1)
      DIMENSION D(12),DEP(21),DATA(21),PV(12)
      IC=L1+1
      IF(IC.GE.12) GO TO 400
      Z=100.
      VEL=PV(L1)
      RHO=1.02
      STRESS=0.003*RHO*VEL**2
      USTAR=SQRT(STRESS/RHO)
      IF(VEL.EQ.0.) GO TO 300
      ZNOT=Z/EXP(0.4*ABS(VEL)/USTAR)
      DO 1010 I10=IC,11
      Z=(DEPTH-D(I10))*100.
      PV(I10)=(USTAR/0.4)*ALOG(Z/ZNOT)
      IF(VEL.LT.0.) PV(I10)=-PV(I10)
 1010 CONTINUE
      GO TO 400
  300 DO 320 LC1 = IC,12
      PV (LC1) = DATA (N)
  320 CONTINUE
  400 PV(12)=0.0
  500 RETURN
      END
```

```
C-----------------------------------------------------------------------------
C      SUBROUTINE: SALINITY AND SIGMA-T COMPUTATIONS
C      THIS SUBROUTINE COMPUTES SALINITY AND SIGMA-T VALUES FROM
C      FROM CONDUCTIVITY AND TEMPERATURE ACCORDING TO COX,R.A.,F.CULKIN, AND
C      J.P.RILEY. 1967. 'THE ELECTRICAL CONDUCTIVITY/CHLORINITY RELATIONSHIP
C      IN NATURAL SEA WATER'. DEEP-SEA RESEARCH 14:203-220.
C-----------------------------------------------------------------------------
       SUBROUTINE SALT (COND,TEMP,SAL,SIGT,N)
       DIMENSION COND(21),TEMP(21),SAL(21),SIGT(21)
       DO 1010 I10=1,N
       IF(TEMP(I10).EQ.999.) GO TO 50
       IF(COND(I10).EQ.999.) GO TO 50
       GO TO 100
   50  SAL(I10)=999.0
       SIGT(I10)=999.
       GO TO 1010
  100  A=TEMP(I10)*0.0297175
       B=(TEMP(I10)*TEMP(I10)*1.5551)*0.0001
       C=-(0.789*TEMP(I10)**3)*0.000001
       CT35=29.03916*(1.0+A+B+C)
       RT=COND(I10)/CT35
       DEL15=0.00001*RT*(RT-1.)*(TEMP(I10)-15.)*(96.7-72.*RT+37.3*RT**2-(
      *0.63+0.21*RT**2)*(TEMP(I10)-15.))
       R15=DEL15+RT
       SAL(I10)=-0.08996+28.2972*R15+12.80832*R15**2-10.67869*R15**3+5.98
      *624*R15**4-1.32311*R15**5
       A=TEMP(I10)*1.0E-3*(4.7867-0.098185*TEMP(I10)+0.0010843*TEMP(I10)*
      **2)
       B=TEMP(I10)*1.0E-6*(18.03-0.8164*TEMP(I10)+0.01667*TEMP(I10)**2)
       Q=-((TEMP(I10)-3.98)**2*(TEMP(I10)+283.))/(503.57*(TEMP(I10)+67.26
      *))
       SIGO=-0.093+0.8149*SAL(I10)-4.82E-4*SAL(I10)**2+6.8E-6*SAL(I10)**3
       SIGT(I10)=Q+(SIGO+0.1324)*(1.0-A+B*(SIGO-0.1324))
 1010  CONTINUE
       RETURN
       END
```

```
C-----------------------------------------------------------------------------
C      SUBROUTINE:VECTORIAL COMPUTATION OF CURRENT VELOCITY DATA
C      THIS SUBROUTINE COMPUTES THE U AND V COMPONENTS FROM CURRENT
C      VELOCITY AND DIRECTION DATA.
C-----------------------------------------------------------------------------
       SUBROUTINE VECTOR(PV,VEL1,L1,UVEC,VVEC,ALPHA)
       DIMENSION PV(12),VEL1(12),UVEC(12),VVEC(12)
       DO 1010 I10=1,12
       XMAG = PV (I10)
       BETA=VEL1(I10)-ALPHA
       PI=3.1415926
       BETA=BETA*PI/180.
       UVEC(I10)=XMAG*COS(BETA)
       VVEC(I10)=XMAG*SIN(BETA)
       IF(UVEC(I10).LT.0.0.AND.UVEC(I10).GT.-1.0) UVEC(I10)=0.
       IF(VVEC(I10).LT.0.0.AND.VVEC(I10).GT.-1.0) VVEC(I10)=0.
 1010  CONTINUE
       RETURN
       END
```

```
C-----------------------------------------------------------------------
C      SUBROUTINE: EXTRAPOLATION OF DATA PROFILES OUTSIDE THE SPLINE RANGE
C      THIS SUBROUTINE EXTRAPOLATES THE DATA PROFILES OUTSIDE OF THE RANGE
C      DEFINED BY THE TOP AND BOTTOM DATA MEASUREMENTS ASSUMING THAT THE
C      FIRST DERIVATIVE AT THE END POINTS REMAINS CONSTANT TO THE SURFACE AND
C      BOTTOM RESPECTIVELY.  IN THE CASE OF VELOCITY MEASUREMENTS EXTRAPOLATION
C      IN THE BOTTOM BOUNDARY LAYER, WHEN NEEDED, IS ENDED 1 M ABOVE THE
C      BOTTOM WHERE SUBROUTINE LOGFIT TAKES OVER.
C-----------------------------------------------------------------------
       SUBROUTINE OUTSD (X1,Y1,X,Y,SM)
       SM = (Y - Y1) / (X - X1)
       RETURN
       END
```

```
C-----------------------------------------------------------------------
C      SUBROUTINE: COMPUTATION OF CUBIC SPLINE EQUATIONS
C      THIS SUBROUTINE IS USED TO DETERMINE INTERPOLATED DATA VALUES FROM
C      THE CUBIC SPLINE FIT GIVEN BY PENNINGTON, R.H. 1970. 'INTRODUCTORY
C      COMPUTER METHODS AND NUMERICAL ANALYSIS.' SECOND EDITION. THE MCMILLAN
C      COMPANY: 445-452.
C-----------------------------------------------------------------------
       SUBROUTINE SPLIN(X,Y,M,C,XINT,YINT)
       DIMENSION X(1),Y(1),C(4,1)
       IF(XINT-X(1))7,1,2
     1 YINT=Y(1)
       RETURN
     2 K=1
     3 IF(XINT-X(K+1))6,4,5
     4 YINT=Y(K+1)
       RETURN
     5 K=K+1
       IF(M-K)7,7,3
     6 FAC1=X(K+1)-XINT
       FAC2=XINT-X(K)
       FAC3=FAC1*FAC1
       FAC4=FAC2*FAC2
       YINT=FAC1*(C(1,K)*FAC3+C(3,K))
       YINT=YINT+FAC2*(C(2,K)*FAC4+C(4,K))
       RETURN
     7 YINT=0.0
       RETURN
       END
```

```
C-----------------------------------------------------------------------
C     SUBROUTINE: COMPUTATION OF CONSTANTS USED IN SPLINE INTERPOLATION
C     THIS SUBROUTINE COMPUTES THE FOUR CONSTANTS NEEDED FOR EACH
C     DEPTH-INCREMENT IN FITTING A CUBIC SPLINE TO THE MEASURED DATA.
C     (PENNINGTON, R.H. 1970. 'INTRODUCTORY COMPUTER METHODS AND NUMERICAL
C     ANALYSIS.' SECOND EDITION. THE MCMILLAN COMPANY: 445-452.)
C-----------------------------------------------------------------------
      SUBROUTINE SPLIC (X, Y, M, C)
      DIMENSION X(1),Y(1),C(4,21),D(21),P(21),E(21),A(21,3),B(21),Z(21)
      MM = M - 1
      DO 2 K = 1,MM
      D (K) = X(K+1) - X(K)
      P (K) = D(K) / 6.0
    2 E (K) = (Y (K+1) - Y (K)) / D (K)
      DO 3 K = 2,MM
    3 B (K) = E (K) - E (K-1)
      A (1,2) = 0.0
      A (1,3) = 0.0
      A (2,3) = P (2) - P (1) * A (1,3)
      A (2,2) = 2.0 * (P (1) + P (2)) - P (1) * A (1,2)
      A (2,3) = A (2,3) / A (2,2)
      B (2) = B (2) / A (2,2)
      DO 4 K = 3,MM
      A (K,2) = 2.0 * (P (K-1) + P (K)) - P (K-1) * A (K-1,3)
      B (K) = B (K) - P(K-1) * B (K-1)
      A (K,3) = P (K) / A (K,2)
    4 B (K) = B (K) / A (K,2)
      Q = D (M-2) / D (M-1)
      A (M,1) = A (M-2,3)
      A (M,2) = D (M-2) - A (M,1) * A (M-1,3)
      B (M) = B (M-2) - A (M,1) * B (M-1)
      Z (M) = B (M) / A (M,2)
      MN = M - 2
      DO 6 I = 1,MN
      K = M - I
    6 Z (K) = B (K) - A (K,3) * Z (K+1)
      Z (1) = -A (1,2) * Z (2) - A (1,3) * Z (3)
      DO 7 K = 1,MM
      Q = 1.0 / (6.0 * D (K))
      C (1,K) = Z (K) * Q
      C (2,K) = Z (K+1) * Q
      C (3,K) = Y (K) / D (K) - Z (K) * P (K)
      C (4,K) = Y (K+1) / D (K) - Z (K+1) * P (K)
    7 CONTINUE
      RETURN
      END
```

Index